BASIC ENGINEERING CALCULATIONS FOR CONTRACTORS

BASIC ENGINEERING CALCULATIONS FOR CONTRACTORS

August W. Domel, Jr., Ph.D.

McGraw-Hill

New York San Francisco Washington, D.C. Auckland Bogota
Caracas Lisbon London Madrid Mexico City Milan
Montreal New Delhi San Juan Singapore
Sydney Tokyo Toronto

Library of Congress Cataloging-in-Publication Data

Domel, August W. (August William), 1960-
Basic engineering calculations for contractors / August W. Domel.
p. cm.
Includes index.
ISBN 0-07-018002-4
1. House construction. 2. Structural engineering. I. Title.
TH4812.D6597 1996
690-DC20 96-33180
CIP

McGraw-Hill

*A Division of The **McGraw-Hill** Companies*

5 6 7 8 9 0 QSR/QSR 6 5 4

ISBN 0-07-018002-4

The sponsoring editor for this book was April Nolan, the editing supervisor was Scott Amerman, and the production supervisor was Don Schmidt. It was set in Garamond by Lisa Lehocky.

Printed and bound by Quebecor

McGraw-Hill books are available at special quantity discounts to use as premiums and sales promotions, or for use in corporate training programs. For more information, please write the Director of Special Sales, McGraw-Hill, 11 West 19th Street, New York, NY 10011. Or contact your local bookstore.

Dedication

I dedicate this book to my mentors throughout my life who took the time to make a difference. In particular, I sincerely thank the following individuals for their efforts, encouragement, and generosity:

John Garvey, Mt. Greenwood Little League

John Flynn, St. Christina Football

Dan Bruch and Don Bruch, Industrial Sling

Kurt S. Tannenwald, Assistant Chief Bridge Engineer for the City of Chicago

Patrick Mazza, Principal of Patrick Mazza & Associates

Contents

Preface

The primary goal of this book is to present the fundamentals of the technical aspects of residential construction. Many books address understanding residential blue prints, becoming a general contractor and managing projects. Presently no book adequately presents the technical side of residential construction. The purpose of this book is to fill that void.

To maximize the benefit of this book it was written to address a wide audience. First, and foremost its focus was to be a valuable resource for the contractor with limited technical training and maximum desire to understand all facets of their work product. It was also written with the engineer and architect in mind. These professions typically have limited training in the analysis and design of residential structures.

Writing for these audiences was facilitated by dividing each chapter into two parts; "Introduction" and "Advanced Discussion." The Introduction portion was written for the contractor and engineer/architect with limited technical background in residential design. Information is provided in a form that will allow the reader to grasp the basics of design of a residential structure. This knowledge should translate into a reduction in field mistakes, a better understanding of alternatives for rectifying field problems and even the providing of a critique of the design drawings prior to construction. The information available in the Introduction portion is thorough enough that nearly an entire residential design is possible.

The Advanced Discussion portion of a chapter either expands the material in the Introduction portion or addresses a more complex issue of the subject topic. The Advanced Discussion is written in a manner that reduces the complex issues to a usable form.

Although it would be useful to read this book from cover to cover, it is not necessary. The book can be used as a technical resource when specific issues arise. Although not achieved in all cases, the intent is for each chapter to stand alone; limited reference is made to other chapters.

The technical fundamentals of residential construction require more than discussion on stresses and strains. Consideration must be given to material properties as well as determination of the loads that cause the stresses and strains.

The topics discussed in this book are organized into six different categories:

- Materials
- Loads
- Analysis
- Wood Members
- Steel Members
- Concrete Members

The categories are subdivided into several chapters. A brief overview of each chapter follows.

Chapter 1–Fundamentals

The fundamentals presented are concepts taught in introductory college engineering courses. These concepts sometimes can become quite complex. This chapter avoids the complexity by only providing a brief overview of the topics of interest to avoid intimidating the reader. This is the only chapter that is needed as background to other chapters. Failure to understand the concepts in this chapter will in no way prevent the use of the other valuable material herein, but it might prevent its full understanding.

The fundamentals addressed in this chapter include:

- Determination of support forces, bending moments and shear.
- Properties of materials
- Dimensional properties
- Material strengths
- Units

Chapter 2–Wood

This the first of four chapters on materials used in residential construction. Wood is an extremely versatile material that can be used for structural purposes (beams and columns) as well as functional and decorative purposes (cabinets and trim).

Topics of interest presented include discussion on species of lumber, grain orientation and material properties. The strength of wood is dependent on number of factors, including the size of the member, its moisture content, the duration of loading, the orientation of the grain, etc. These factors are presented and the corresponding reduction and increases in allowable strength are presented. Information is also provided on nails, plywood and lumber grading.

The Advanced Discussion is devoted to providing detailed information on the load carrying abilities of plywood. Plywood is used for roof sheathing and for the floor surface. Analysis examples are provided for plywood that is supported on 48-in., 24-in. and 16-in. centers.

Chapter 3–Steel

The proportion of steel used in residential construction is small when compared to the volume of other material used in the structure. The steel members, although small in amount, carry loads of significant magnitude. Steel is used for beams that support the ends of wood floor joists and also for the columns that support the beams.

This chapter provides information on the cross-sectional shapes of steel members that are commercially available. The method of specifying steel beam sizes and material classifications are presented.

Steel beams and columns are individual elements that form a rigid structure when connected to each other. In residential construction connections are not as sophisticated as those needed in commercial construction. Situations might arise where a more significant connection is required, although this is not the norm.

Connections for steel members are facilitated by bolting or welding. The Advanced Discussion part of this chapter includes discussion on bolt types, method of measuring the tightness of a bolt and types of welded joints.

Chapter 4–Concrete

Concrete is utilized in residential construction for foundations, sidewalks, slabs and driveways. The composition of the concrete mix, types of concrete, and types of tests are discussed in this chapter. Common defects and distress in concrete are also discussed. The Advanced Discussion part of the chapter includes information on reinforcing concrete with steel bars, prestressed concrete and concrete additives. These concepts are not widely used in residential construction but the advantages they provide might justify their use in the future.

Chapter 5–Masonry

This chapter is the last of the chapters on materials. Masonry can be used for structural purposes (retaining walls and basement walls) or as a decorative element (facade, fences or sidewalks). This chapter provides information on how bricks are categorized and manufactured, and the pertinent specifications.

Masonry walls can be vulnerable to failure caused by wind forces. This type of wall is most vulnerable to collapse from wind forces immediately after it is constructed. At that time, the mortar has not gained sufficient strength to bond the bricks into one solid unit. The masonry wall is a pile of bricks until the mortar has set, and tall walls can topple from wind forces. The Advanced Discussion provides guidelines for determining when a wall should be braced and a detail of how this bracing can be accomplished.

Chapter 6–Codes, Specifications and Recommended Practices

Many a builder has heard the phrase "it does not meet Code." This chapter provides a list of the codes used in residential construction and the entities that developed these codes.

Codes also reference other codes, specifications and documents. The more common of these referenced items and their producers are given as well as a brief description of the document. Documents are divided into categories for easy reference. For readers interested in more detailed information, the Advanced Discussion provides information on more technical specifications and books.

Chapter 7–Dead and Live Loads

Buildings must be designed to withstand a variety of loads. Some areas of the country require consideration of seismic forces and other areas must contend with high wind forces. But regardless of locale, all buildings must be designed to support their own weight (called *dead load*) and the loads from occupants (called *live load*).

This chapter provides the design live loads for a variety of structures including residential construction. Design live loads for residential structures are presented for various locations within the structure. Also included are the weights of material for various components of the structure.

The building codes recognize that it is highly unlikely that the design live load will cover the entire floor at the same moment in time. For this reason the code allows a reduction in live load when large floor areas are being designed. This concept and the corresponding calculations are provided in the Advanced Discussion.

Chapter 8–Roof Loads

The loads on the roof consist of the weight of the shingles, rafters, live loads, snow loads and wind loads. Dead loads are presented in Chapter 7 and wind loads are presented in Chapter 9. This chapter discusses roof live loads and snow loads.

The slope of the roof has a relationship to the magnitude of the snow load. The magnitude of the snow load is also related to the local terrain and geographical location of the structure. These topics are discussed in detail in this chapter. The Advanced Discussion part of the chapter presents the loading on roofs when drifting of the snow is possible.

Chapter 9–Wind Loads

Structures must be able to withstand unpredictable and possibly extremely powerful wind forces. Destruction caused by the wind forces and common failure modes are discussed in this chapter. The Advanced Discussion provides details on how to calculate the wind pressures for which the structure should be designed in a given locale. Sample calculations are provided.

Chapter 10–Deflections

The strength of a beam has obvious importance in relation to the load-carrying capacity of a beam. A beam that has sufficient strength might not be acceptable if it has excessive deflections. This chapter provides the equations to calculate the deflections for a variety of beams with a variety of loading conditions. The Advanced Discussion portion presents the results for beams that span over multiple supports.

Chapter 11–Forces

As noted in Chapter 1, bending moments and shear forces produce stresses in the beam. These stresses must be calculated to determine if the strength of the beam is in excess of the applied forces. This chapter presents the equations for determining bending moments and shears for the same load cases presented in Chapter 10. These items are used to determine the stress in a member.

Chapter 12–Wood Joists

This chapter is one of three chapters devoted to wood members. The loads on floors and ceilings are carried by shallow, closely spaced structural members called *joists*. Wood joists range in size from 2 × 6s to 2 × 16s. This chapter provides the design of wood joists to satisfy deflection, bending moment stresses and shear stress requirements.

Tables in this chapter show allowable stresses for several wood species. They also display the proper joist size for a given loading on a given length of beam. The Advanced Discussion includes a detailed design example.

Chapter 13—Wood Trusses

Trusses provide one of the most economical methods of carrying loads. These structural members can span in excess of 40 ft. and trusses are available in a wide variety of configurations. The most popular trusses and the length they span are presented in this chapter. The Advanced Discussion displays tables that compare the forces in each truss member as the pitch of the roof changes. These data enable the reader to understand the need for an increase in design forces as the roof becomes flatter.

Chapter 14—Wood Headers

A structure is of little value if it does not have openings in the frame for windows and doors. These openings interrupt the load path that normally goes straight down to the foundation below. Therefore, provisions must be made to transfer the loads directly above the opening to the sides of the opening so that the loads can be transferred to the foundation. This objective is accomplished by using a beam above the opening called a *header*.

This chapter presents the design procedures for a header. Discussion includes information on typical lumber headers and glued laminated beams.

Chapter 15—Steel Beams

This chapter discusses the use of steel beams to support floor and wall loads. Tables provide load capacities of steel beams commonly used in residential construction. Calculations to determine the load to be carried by a steel beam are presented in the design example.

The Advanced Discussion portion of this chapter includes a detailed beam design example. The results are compared to the tabular results presented earlier in the chapter.

Chapter 16—Steel Lintels

As discussed in Chapter 14, a functional house requires openings in the structure to provide exits, ventilation and light. When the exterior of a building is constructed of brick, steel members are required to carry the heavy masonry loads to the sides of the opening. This chapter presents the geometric properties of steel angles and the height of brick that each angle can support. The Advanced Discussion portion of this chapter includes two detailed lintel design examples.

Chapter 17–Concrete Slabs

Concrete slabs, when properly placed, have long service lives with minimal maintenance. The drawback to using concrete is its susceptibility to cracking. This chapter discusses the common problems that lead to cracking in slabs. The Advanced Discussion of this chapter discusses proper curing techniques to reduce the possibility of shrinkage cracking. The concept of crack control by using reinforcing steel is also discovered.

Chapter 18–Concrete Footings

Concrete walls must transfer their loads into the underlying soil. If the loads on the bottom of the wall are in excess of the bearing capacity of the soil, failure or unsightly distress will result. This problem is avoided by enlarging the bottom of the concrete wall, thereby reducing the applied pressure to the soil. This chapter presents the common types of wall footings and their loads.

Chapter 19–Concrete Walls

Concrete walls are used to retain the soil so that the structure can have a basement or a crawl space. These walls must be strong enough to withstand the earth pressures from the soil, particularly when the soil is wet. This chapter discovers the fundamentals of the design of concrete walls used in residential construction.

Acknowledgments

The desktop publishing for this text was done by Lisa Lehocky of Round Lake Park, Illinois. The final work product was accomplished from literally hundreds and hundreds of pages filled with rough sketches and sloppy handwriting. Without a complaint, she turned it into a well organized and readable work product. She has done a fantastic job and I sincerely thank her.

Roman Szczesniak of Woodridge, Illinois, reviewed the text for accuracy. More importantly, he provided the author with a resource to discuss many technical aspects of the book during its development stages. His thorough understanding of both the technical and practical sides of structural engineering is impressive, to say the least. I thank him for his valuable input.

Disclaimer

This text was written for the purpose of introducing contractors and other individuals to the technical side of residential construction. Reading this text by no means makes one a qualified designer, engineer or architect. It will however make the reader better informed to participate in the design process, fixing field problems and understanding the technical side of their work product.

The information contained herein was written specifically for single-family residential construction. Discussions, calculations, figures, and tables were prepared with this structure in mind. It is not advisable that the principles in this text be used for any other type of structure.

This text was reviewed for accuracy and for technical content. A great deal of effort was spent on trying to produce an error-free text. In reality, such a goal is unattainable. Therefore, I caution the reader of the possibility of mistakes in the book. I encourage that the reader notify the publisher if any mistakes are found. The author assumes no responsibility for anyone's use of this text and its principles.

BASIC ENGINEERING CALCULATIONS FOR CONTRACTORS

1

Fundamentals

Introduction

This book caters to those with limited technical backgrounds in the analysis and design of residential structures. This chapter presents the fundamental principles that will assist in understanding the remainder of this book. Although these fundamentals are helpful in understanding the principles, it in no way precludes the reader from utilizing the rest of the text. Fundamentals are essential to some but boring and tedious to others. If the reader has no desire to learn the basics, then he or she should move on to the next chapter. However, for him or her to get the full impact of this book, some of the principles provided in the remainder of this chapter will be of value.

The discussion presented in this chapter includes:

- Introductory material
- Bending moments
- Shears
- Support reactions
- Modulus of elasticity
- Units consistency

Preliminary questions that arise when the design of a structural member is considered include how the load is applied and the support considerations.

In residential construction the loads can be in the form of a concentrated load or a uniformly distributed load. A concentrated load has its entire force located at one point. An example of a concentrated load is a refrigerator located in a span of a floor joist. In this example the load is applied over a very small area and is considered concentrated.

The uniformly distributed load is the complete opposite of a concentrated load. A uniformly distributed load is spread evenly across the structural member. An example of this condition would be snow on a roof. If the snow is the same depth at all locations, it will produce a uniform load on the structural members that support it. Figure 1-1 shows a schematic of both a concentrated load and a uniformly distributed load.

The method in which the structural member is supported will effect the stresses that will develop. Types of supports fall into two groups, fixed and simply supported.

Figure 1-2a shows a cantilever beam, which is a beam supported at one end. The support end of cantilever beam is considered to be a "fixed" support. A fixed support prevents the beam from moving down at the supported end. It also prevents the supported end of the beam from rotating.

At the other end of the spectrum is the simple support (also called *simply supported*). This support condition is depicted in Figure 1-2b. The ends of this beam are supported so that there is no downward displacement, but the supports do not prevent rotation.

Figure 1-3a presents the difference between a simple supported and a fixed support. In Figure 1-3a the beam is resting on top of a concrete wall. The wall will support the beam but will not prevent the ends from rotating. Figure 1-3b depicts the shape the beam will take when loads are applied.

In Figure 1-3c the ends of the beam are embedded in the wall. This embedment will prevent the ends of the beam from rotating. Figure 1-3d depicts the shape the beam will take when it is loaded. Therefore the beam depicted in Figures 1-3a and 1-3b are simply supported, and those in Figure 1-3c and 1-3d have fixed supports.

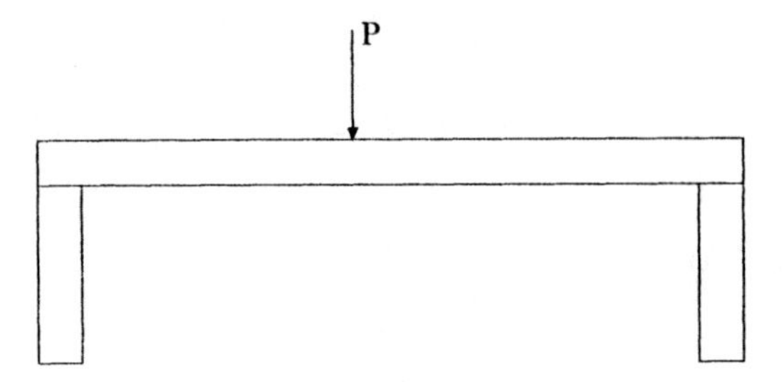

1a: Concentrated load at midspan of beam

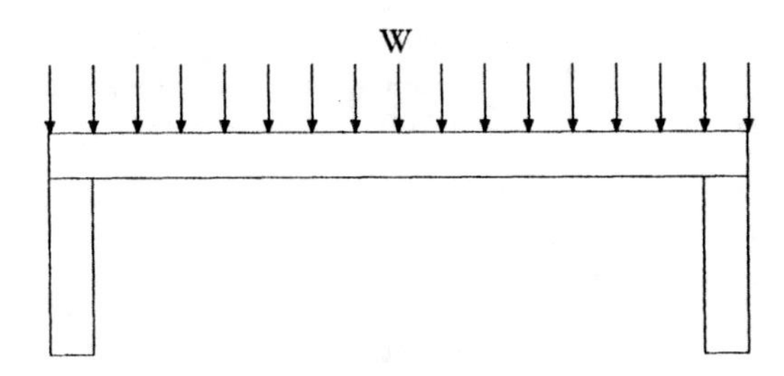

1b: Uniformly distributed load across length of beam

Figure 1-1—Types of beam loads

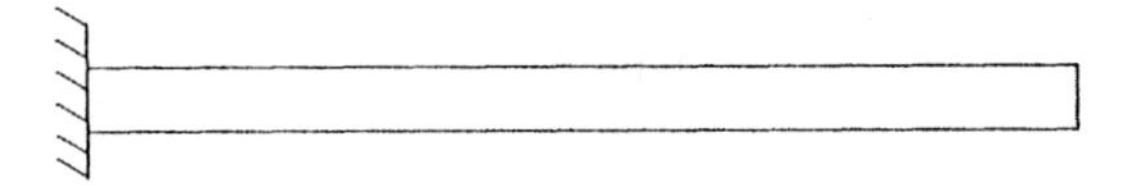

2a: Beam with fixed end

2b: Beam with simple supported ends

Figure 1-2—Support conditions

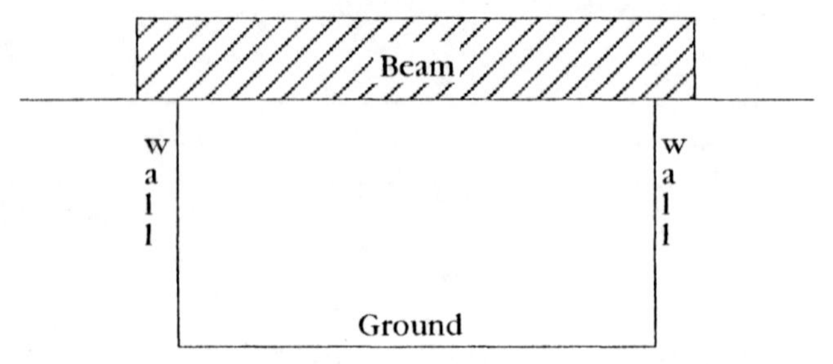

3a: Simply supported beam

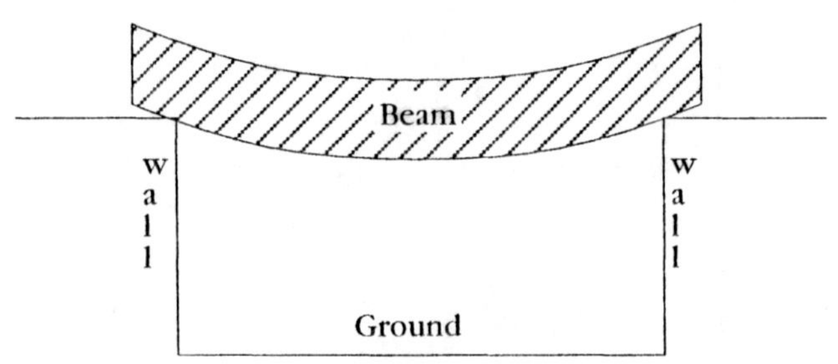

3b: Deflection of simply supported beam

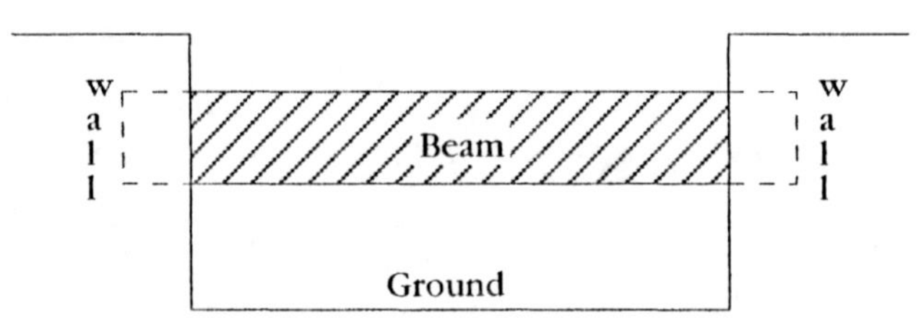

3c: Beam with fixed end supports

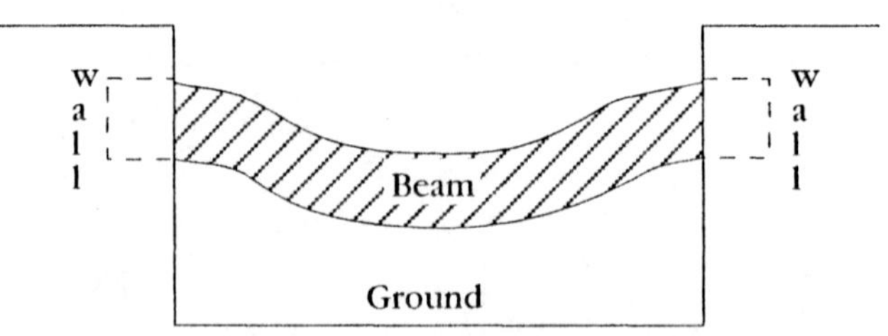

3d: Deflections of beam with fixed end supports

Figure 1-3—Examples of support conditions

In reality a beam support is neither fixed nor simply supported. On one end of the spectrum a beam is never totally free to rotate and on the other end, it is not possible to connect a beam where rotation is totally restrained. One might get the impression that the nailing of joist ends would provide a fixed support. This end restraint is minimal and not nearly enough to prevent rotation.

A load on a beam will cause deflection and stresses within the beam. Understanding the concept of deflections is not difficult; however, the concept of the induced stresses is not as straight forward. For analysis purposes, the stresses in beams can be divided into bending moment stresses and shear stresses.

When a beam is loaded it will take a curved shape, as shown in Figure 1-4. This curvature induces a bending moment. Figure 1-4b shows the same beam cut at midspan. As shown in this figure the bending effect can be reproduced by adding a twisting motion at the cut. This twisting motion is called a *bending moment.* Chapter 11 discusses in detail how to determine the bending moment for a variety of loading conditions. Calculation of the stresses that result from the bending moments are discussed in this chapter.

Regardless of how the bending moment is produced, the resulting stress is calculated using the following equation:

$$\text{Bending stress} = Md/2I$$

Where M is the bending moment, d is the height of the member and I is the moment of inertia. The value of M is determined using the information in Chapter 11, d is easily known and I is the value of the moment of inertia, a geometric property of the section. For a rectangular section the value of I is determined as:

$$I = \frac{(\text{width})(\text{height})^3}{12}$$

For example, if a beam is 10 in. high and 1.5 in. wide, the moment of inertia is:

$$I = \frac{(1.5)(10 \times 10 \times 10)}{12}$$

$$= 125 \text{ in.}^4$$

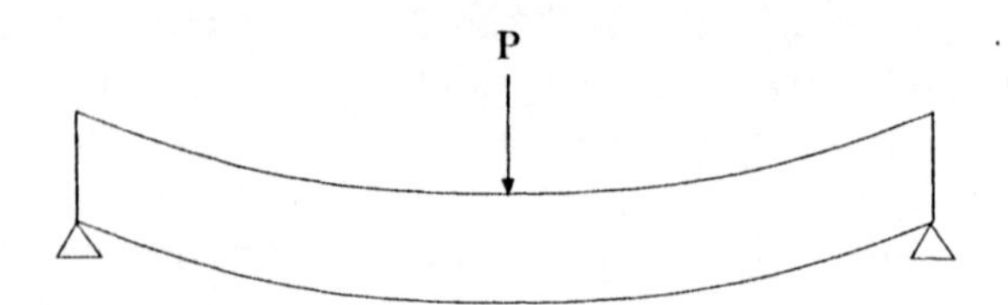

4a. Deflected beam

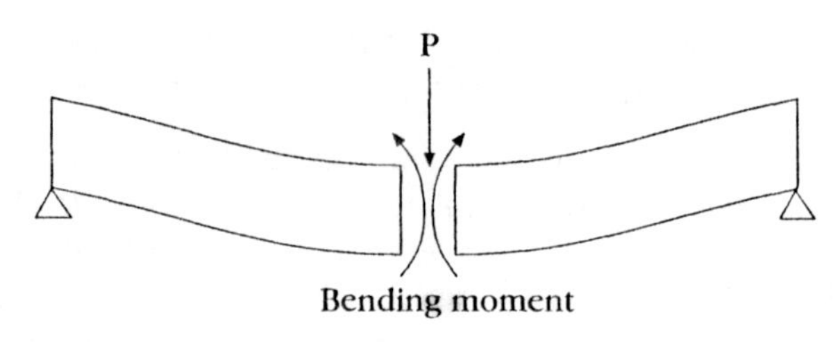

4b. Bending moments for a deflected beam

Figure 1-4—Bending moment in beam

If the section is not rectangular, the calculation is more complicated. These values can be obtained from handbooks on this subject.

In summary, if the height of the member is known, the bending moment is known and the moment of inertia is determined, then the bending stress can easily be calculated.

Note that the bending stress is compressive on the top of the beam and tensile on the bottom of the beam for the beam shown in Figure 1-5. As the beam bends to the configuration shown in the figure, the top fibers of the beam will shorten (i.e., compression) and the bottom fibers will expand (i.e., tension). If the loading caused the beam to bend upward, the relationship would be reversed. It is important to determine which face has tensile forces and which has compressive. Some materials such as concrete have excellent compressive resisting qualities and terrible tensile resisting qualities.

The other stress to consider in beam design is shear stress. Shear stress is the propensity of the beam to have one face slip past

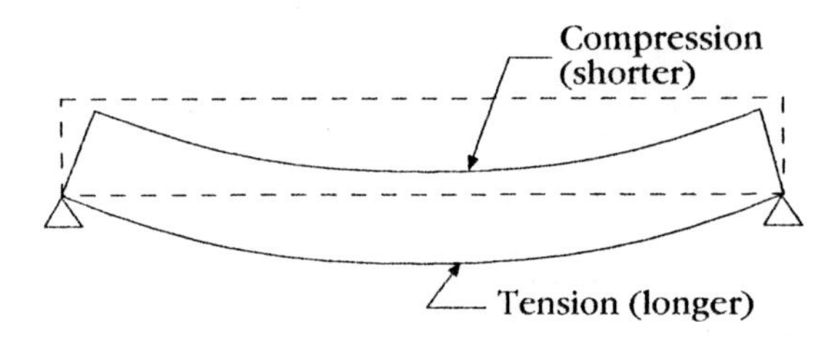

Figure 1-5—Tensile and compressive forces in a beam

another, as shown in Figure 1-6. The shear stresses can be determined with the following equation:

Shear stress = 1.5V/b

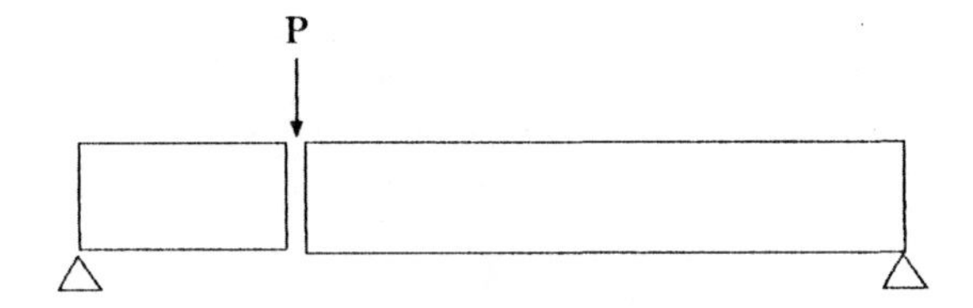

Figure 1-6—Shearing of beam

The determination of the shear force, V, is discussed in detail in Chapter 11. The other two variables b and d are the width and height of the member respectively.

In addition to determining the stresses, the forces on the supports need to be known. Forces on the supports, called *end reactions*, are dependent on the method used to support the beam and the location of the loads. Figure 1-7 depicts a simply supported beam with a concentrated load at midspan. It is easy to see that the force would be carried equally by both supports. The situation gets slightly more complicated when the load is not centered. Figure 1-7b depicts a simply supported beam with a load located closer to the left end. By proportions, the left support will experience a greater share of the applied load.

Deflections of beams with various load conditions are discussed in detail in Chapter 12. Regardless of the loading, one factor common to all of the equations is the value E. This factor E, called the modulus of elasticity, is a numerator in the deflection equation. The modulus of elasticity is a measure of how much spring the material has. A member with a high modulus of elasticity will

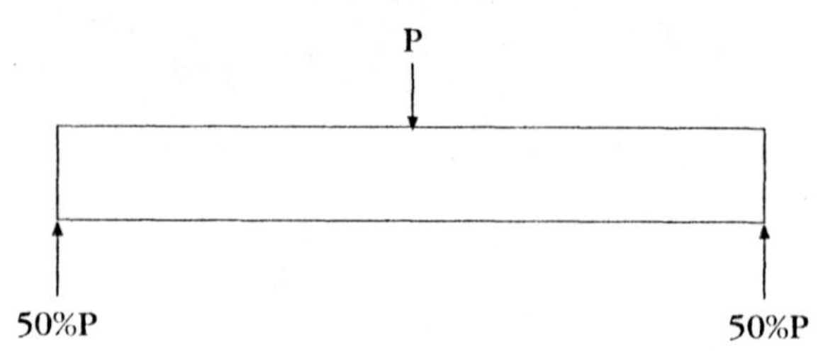

7a. End reactions with load at midspan

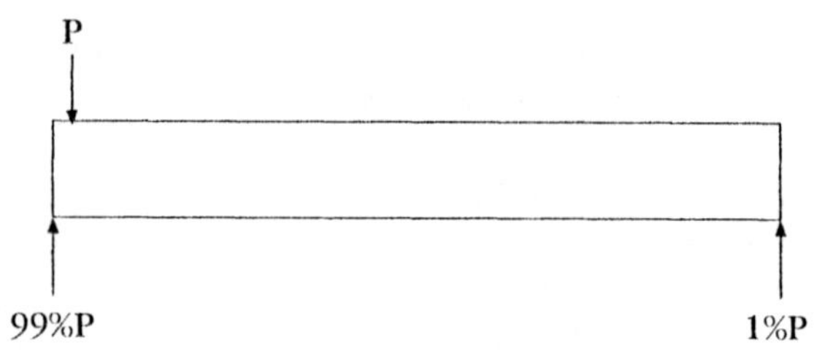

7b. End reactions with load near left support

Figure 1-7–Support reactions

deflect more than a material with a small E under the same loading conditions.

When one reads through this book, it is important to understand the various units used and to keep their use consistent. If one compares an answer in feet to an answer in inches, the results are worthless. The following is a reference to common units.

Length Units

Feet: ft
Inches: in.

Force Units

Pounds: lb

Stress Units

Pounds per square inch: psi or $lb/in.^2$
Pounds per square foot: psf or lb/ft^2

Bending Moments

Inch-pounds: in.-lb
Foot-pound: ft-lb

Materials

It would be rare that a building, particularly a residential structure, would be constructed using only one material for the load-carrying members. Houses are usually constructed with wood walls and floors, steel beams for heavily loaded spans and concrete or masonry for the foundation. With advances in technology it is possible that in the future these materials might also assume the rules of their counterpart materials.

It would not be surprising that some materials such as plastics and fiberglass could be used as structural members in the future. This text is limited to presenting the following materials:

Chapter 2–Wood

Chapter 3–Steel

Chapter 4–Concrete

Chapter 5–Masonry

Each of these materials has their weak points and strong points. Wood is relatively cheap and easy to install but has limited strength. Steel has significant strength but is heavy and requires more sophisticated connections. Concrete is durable but is susceptible to cracking. Masonry is also durable but is limited to use as a vertical member.

The following four chapters provide the reader with a working knowledge of the materials used to resist applied loads. Where applicable, the strengths of the member and methods of connections are provided.

2

Wood

Introduction

More than two out of every three houses are constructed of wood. Wood is a convenient material to work with since it can be easily modified in size by field cutting, is relatively light in weight and is readily available in a wide array of sizes.

One characteristic of wood that is different from other materials is that wood is anisotropic. Anisotropic materials exhibit different physical properties in each of the three principal directions. Review of Figure 2-1 shows that a piece of wood can be sawn from a tree with the rings of the tree having different orientations. The orientation of the ring will have a direct relationship to the strength of the member.

As is well known, the number of rings in the cross section of the base of a tree can be used to determine the age of the tree. This method does not always give an accurate age since temporary significant changes in weather may produce more than one ring or make adjacent rings indistinguishable.

The rings in a tree are composed of two parts, the earlywood portion and the latewood portion. Earlywood, which is the part of the ring that develops in the beginning of the growth season, is softer and less dense than latewood. Latewood, which grows in the later part of growing season, is denser than the earlywood.

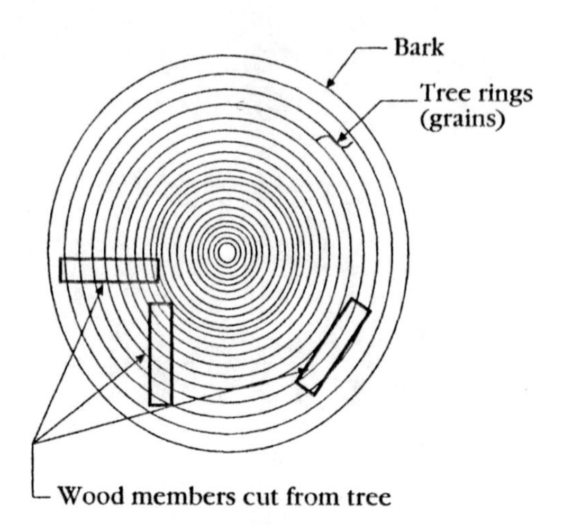

Figure 2-1—Wood members cut from tree with different grain orientation

Typically the greater the percentage of latewood, the stronger the wood.

Wood is divided into two groups, hardwoods and softwoods. This division is misleading since not all of the hardwoods or softwoods are actually hardwood or softwood, respectively. One method to roughly categorize woods is to assume that trees with cones and pine needles are softwoods and those with leaves that fall off in the winter are hardwoods.

Some of the trees that are classified as hardwoods are:

- Oak
- Ash
- Elm
- Walnut
- Hickory

Some typical softwoods are:

- Fir
- Pine
- Cedar

- Spruce
- Hemlock

Although it might appear that hardwoods would be the material of choice for house construction, it is not the case. Hardwoods are more often used for products that have fine finishes such as trim, surface flooring and furniture. Softwoods are used for framing, formwork and sheathing.

The orientation of the grain of the wood will have an impact on the member's strength. A member is strongest when loaded in a direction that is perpendicular to the grain of the wood (see Figure 2-2a). As a corollary, wood is weakest when loaded parallel or not completely perpendicular to the direction of the grain (see Figure 2-2b).

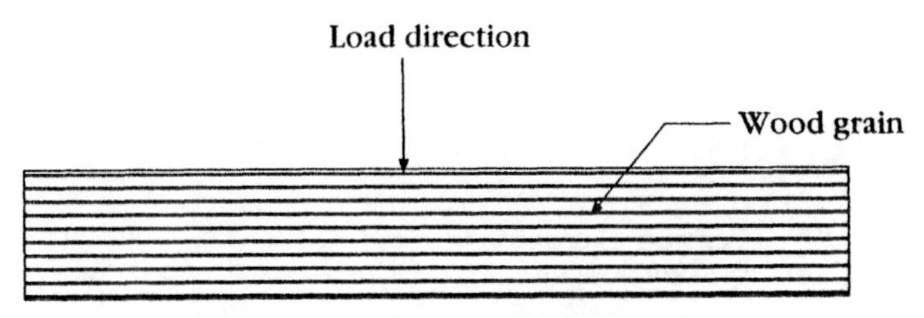

2a. Wood grains perpendicular to load direction

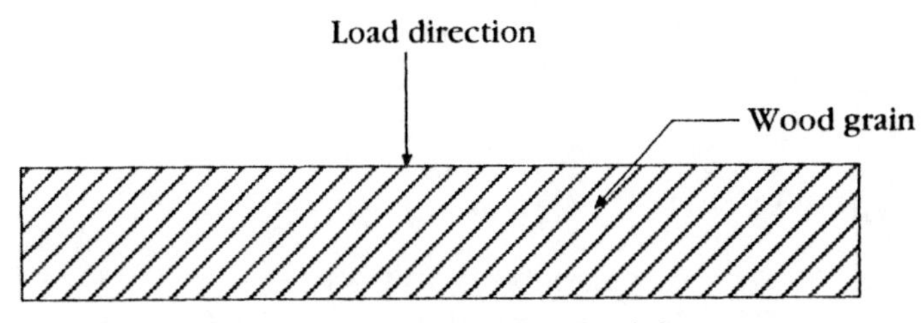

2b. Wood grains not perpendicular to load direction

Figure 2-2–Orientation of wood grain

The physical characteristics of wood that are of interest from a residential design point of view are:

Bending, F_b

Shear, F_v

Modulus of Elasticity, E

Bending stresses in a member occur from the downward bowing of the member. This loading condition (see Figure 2-3) produces tension in the bottom fibers of the wood member. The allowable bending stresses vary depending on the species of wood and the quality of the particular piece of wood. The following are ranges of permissible bending stresses for several types of wood 2 × 12s, as provided by the National Design Specifications Standard for Wood.

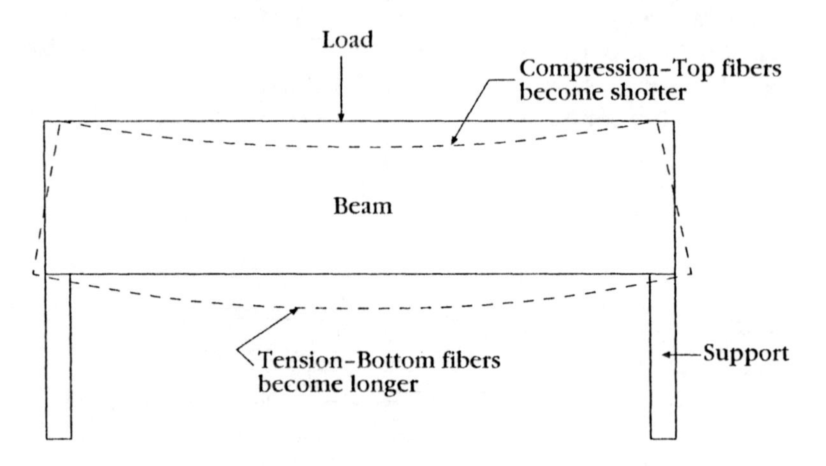

Figure 2-3—Tension and compression in loaded beam

- Spruce-Pine-Fir 750–1300 psi
- Hem-Fir 850–1400 psi
- Douglas Fir 825–1300 psi
- Mixed Oak 800–1150 psi
- Mixed Maple 700–1000 psi
- Western Cedar 700–1000 psi

The shear in a member is the propensity of the two layers of the member to slip in opposite directions (see Figure 2-4). Allowable shears for some types of woods are:

- Spruce-Pine-Fir 70 psi
- Hem-Fir 75 psi

- Douglas Fir 90 psi
- Mixed Oak 85 psi
- Mixed Maple 100 psi
- Western Cedar 75 psi

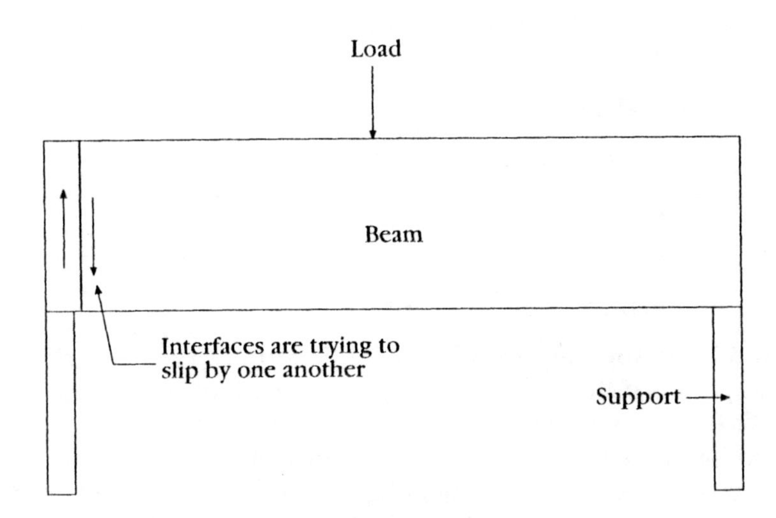

Figure 2-4—Shear in beam

The magnitude of deflection of a member under a given load is inversely proportional to the modulus of elasticity. An increase in the modulus of elasticity by a factor of 2 will reduce the deflections by half. The modulus of elasticity of wood is a function of the density of the wood. The modulus of elasticity for several woods are:

- Spruce-Pine-Fir 1,1400,000–1,300,000 psi
- Hem-Fir 1,300,000–1,600,000 psi
- Douglas Fir 1,200,000–1,400,000 psi
- Mixed Oak 900,000–1,100,000 psi
- Mixed Maple 1,100,000–1,300,000 psi
- Western Cedar 1,100,000 psi

This list shows that the deflection for a quality mixture of hem-fir is about half of a lower-grade oak under the same loading conditions.

The values given are generic for each of the woods and must be adjusted for various factors. Some of these factors increase the allowable stress while others result in a decrease.

The five factors that are applicable to residential wood design that will be discussed are:

- Load duration factor
- Wet service factor
- Beam stability factor
- Beam size factor
- Repetitive member factor

Load Duration Factor: Wood members are capable of carrying heavy loads if the loads are for a short duration. Short-duration loads include impact forces of dropping objects. Loads of medium duration include wind, seismic and snow loads. Loads that are more permanent in nature include floor live loads and dead loads. For short-term loads, the allowable shear and bending stresses can be doubled. For medium-duration loads these allowables can be increased from 15 to 60%. Floor live loads have no allowable increase while permanent dead loads require a 10% reduction in the allowable stresses.

Wet Service Factor: When the moisture content of wood is high (greater than 19%), the properties of the material become less desirable. For this condition, the following reductions might be applicable:

Bending–reduce allowable by 15%

Shear–reduce allowable by 3%

Modulus of elasticity–reduce by 10%

Beam Stability Factor: When the depth of a member is greater than its width, which is always the case for floor joists, lateral support must be provided to prevent the top of the joist from buckling sideways. When the member has continual support at the top of the joist by the floor plywood and with proper X-bracing or bridging, the beam stability factor is 1.0. If the joists are not properly braced against this movement, the reduction could possibly exceed 90% of the allowable. This reduction only applies to the bending stresses.

Beam Size Factor: The allowable stresses in bending were given for a 12-in. deep member. Members of different depths have different allowables. The modification factors are:

2 × 14 Reduce allowable by 10%

2 × 10 Increase allowable by 10%

2 × 8 Increase allowable by 20%

2 × 6 Increase allowable by 30%

2 × 4 Increase allowable by 50%

Repetitive Member Factor: When wood is used for members such as floor joists and roof truss chords, the members are placed side by side in close proximity. Because of this redundancy, an increase in the allowable of 15% can be used if there are at least three members that are joined together as a unit and the spacing does not exceed 24 in.

In summary, the modifiers are as follows:

Modifier	Result
Load duration	Benefit (except dead loads)
Wet service	Detrimental for wet wood
Beam stability	Detrimental for unbraced members
Beam size	Benefit for beams less than 12 in. deep
Repetitive members	Beneficial

Example: Consider a case where a member will be designed to carry snow load. The member is 8 in. deep and has similar members on either side. The beam is fully braced along its length, but the wood is not fully dry (i.e., more than 19% water content). Determine the overall modification factor for bending under these conditions.

Load duration factor = 15% increase in allowable

Wet service factor = 15% decrease

Beam stability factor = 1.0

Beam size factor = 20% increase

Repetitive member factor = 15% increase

$$\text{Overall Modification Factor} = 1.15 \times 0.85 \times 1.0 \times 1.2 \times 1.15 = 1.34$$

Therefore under these conditions the allowable bending stress could be increased by 34%.

The allowable design parameters for a piece of wood is not only a function of the species of wood. The prior growth characteristics of the particular piece of lumber is also important. Obviously a beam with a large knot at midspan will not perform as well as a beam with no knots. In an effort to sort the lumber into various levels of quality a visual rating system places the pieces of wood into categories. The typical categories are:

- Select structural
- Structural No. 1
- Structural No. 2
- Structural No. 3

An individual piece of wood is placed into one of these categories based on the number of knots present, the location of the knots, the slope of the grain of the member and other strength-limiting categories. Select structural is the best and Structural No. 3 is the worst in the categories presented above. Wood used as structural framing for residential conditions would normally be Structural No. 2 and better or Structural No. 1 and better. Each piece of wood used for structural applications must be stamped with an appropriate rating.

It is crucial in understanding the fundamentals of wood to know that a 2 × 12 piece of wood is not 2 in. by 12 in. A 2 × 12 was once 2 × 12, which is called the *nominal size*, when it is in its rough form. Typically wood used for structural framing is dressed lumber. Dressed lumber is the nominal size trimmed to obtain smooth surfaces, which results in decreased dimensions. The actual dimensions of dressed lumber are:

Table 2-1—Lumber Sizes

Type	Width (in.)	Depth
2 × 4	1.5	3.5
2 × 6	1.5	5.5
2 × 8	1.5	7.25
2 × 10	1.5	9.25
2 × 12	1.5	11.25
2 × 14	1.5	13.25
1 × 4	0.75	3.5
1 × 6	0.75	5.5

The variation from the nominal size is important when heights of floors or other dimensions are critical.

Nails are the most common type of connector used to tie the individual elements of a structure to form a rigid unit. The nails used for framing are common and box nails (see Figure 2-5). The size nail is specified by the term *penny*. Penny, designated by the letter d, will determine the diameter and size of the nail. A 6-penny nail, denoted as 6d, is 0.115 inches in diameter and 2 inches long. The dimensions for various size nails are given in the following tables:

Table 2-2—Common Nail Sizes

Size	Length (in.)	Diameter (in.)
4d	1.50	0.10
5d	1.75	0.10
6d	2.00	0.11
7d	2.25	0.11
8d	2.50	0.13
10d	3.00	0.14
12d	3.25	0.14
16d	3.50	0.16

Table 2-3—Box Nail Sizes

Size	Length (in.)	Diameter (in.)
4d	1.50	0.08
5d	1.75	0.08
6d	2.00	0.10
7d	2.25	0.10
8d	2.50	0.11
10d	3.00	0.12
12d	3.25	0.12
16d	3.50	0.13

Other nails used in residential construction industry are (see Figure 2-5):

Double-Headed Nails—Used in an operation where the nail is only temporary and is later removed. Common for nailing formwork.

Finishing Nails—Used when the head of the nail is not to be visible. Common for nailing interior trim.

Roofing Nails—Used to fasten shingles to the wood roof sheathing.

Drywall Nails—Used to attach drywall to the wall studs and ceiling.

Lumber used for constructing frames of houses usually consists of 2 × 4s, 2 × 6s, 2 × 8s, etc. These frames must be covered with plywood to provide a walking surface, rigidity to the frame and a roof surface.

Plywood is constructed of several layers of thin sheets of wood (called *veneer*) which are glued together. Each successive layer of the plywood has its grain oriented perpendicular to the adjacent layers (see Figure 2-6).

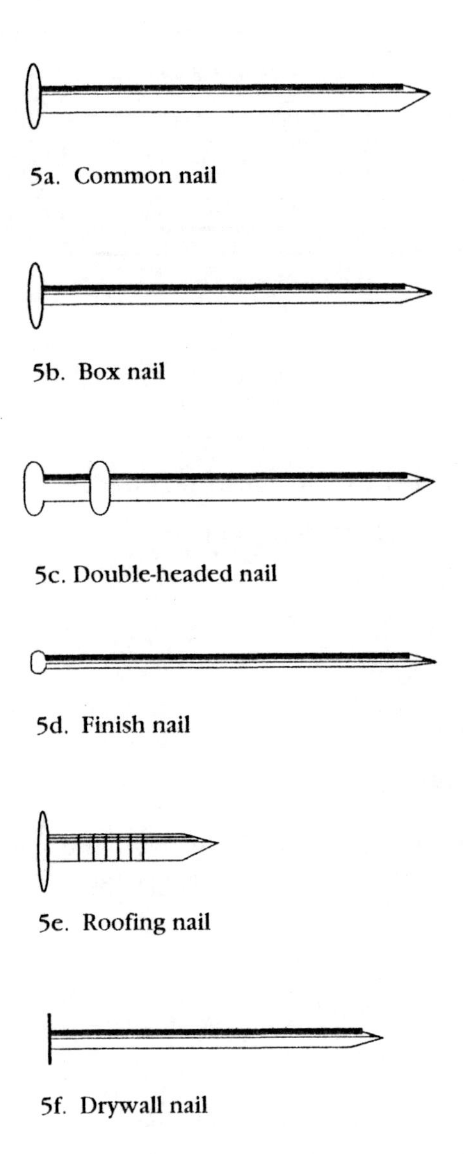

Figure 2-5–Nails used in residential construction

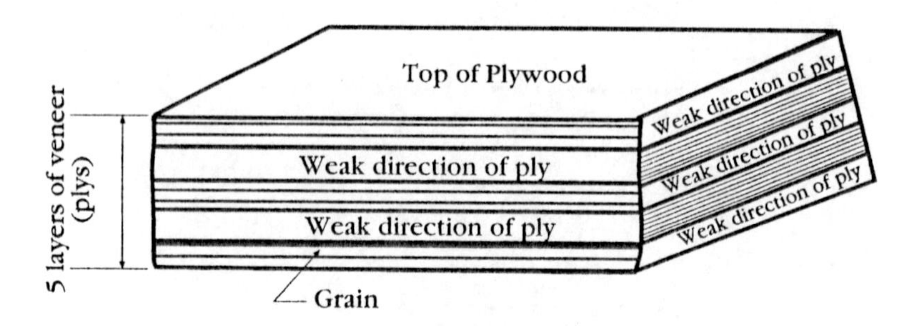

Figure 2-6—Cross section of plywood showing orientation of wood grain

Plywood has many uses in construction including:

- Formwork for placing concrete
- Wall sheathing
- Roof sheathing
- Exterior siding
- Cabinets
- Counter tops

Plywood is distinguished by organizing it into specific categories. Categories rate the wood on various types of characteristics. These characteristics are:

- Exposure durability
- Species of wood
- Veneer quality

Plywood is an essential element for the structural integrity of the house. It must be able to resist the potential problems associated with moisture and humidity. There are cases where the plywood is protected from environmental attack and it is not necessary to consider the effects of moisture and humidity. Plywood is divided into the following four durability classifications:

- Exterior—The composition of plywood in this classification is such that it can withstand exposure to moisture.

- Exposure 1–The composition of plywood in this classification is such that it can withstand temporary exposure to the elements. It is used when exterior plywood is needed only a short period of time, such as temporary exposure to the elements during construction.
- Exposure 2–The composition of plywood in this classification is such that it can only be used for interior of structures. It can be used where some exposure to moisture and humidity is expected.
- Interior–The composition of this plywood is such that it can only be used on the interior in a dry atmosphere.

The next category of interest is the species of wood that is used to construct the plywood. The classification consists of five groups with each group composed of several different species. The groups divide the possible wood choices into species with similar strength characteristics. Group 1 consists of the species of wood that have the highest strength characteristics. Group 5 has the weakest strength characteristics. A partial list of the wood species for the five groups is as follows:

Group 1
- Southern pine
- Larch
- Some douglas fir

Group 2
- Fir
- Western hemlock
- Some douglas fir
- Black maple

Group 3
- Eastern hemlock
- Ponderosa pine
- Redwood
- White spruce

Group 4
- Western red cedar
- Cottonwood

- Aspen
- Eastern white pine

Group 5
- Basswood
- Poplar

The final categorization of plywood is based on the quality of the exposed surface of the plywood. The categories are:

Veneer Grade A–Smooth surface. Limited amount of repairs in the exposed surface are allowed.

Veneer Grade B–Not necessarily a smooth surface. Some splits in exterior face are allowed.

Veneer Grade C–Allows knot holes up to one inch across grain. Some limited splits allowed. If this veneer grade is improved by having limits in the size of cracks and knotholes, it is upgraded to Grade C-plugged.

Veneer Grade D–Allows larger knot holes than other veneer grades.

It is important to note that the direction a piece of plywood is oriented has significant impact on the load-carrying capacity of the plywood.

As previously mentioned, the layers of wood that form the plywood typically have the grain of the wood running in an opposite direction than the layer above. For a three-ply system (see Figure 2-6), two layers have grain running in one direction and only one is in the opposite direction. This arrangement results in a plywood that has significantly more strength when spanning in the direction of the grain of the two plys. For this reason plywood should have the long dimension of the panel spanning across the supports (see Figure 2-7).

The American Plywood Association (APA) provides performance ratings for plywood. An APA performance-rated plywood trademark, which appears stamped on the plywood, contains the following information of importance in residential construction.

1. Grade of panel
2. Allowable span length
3. Thickness of panel

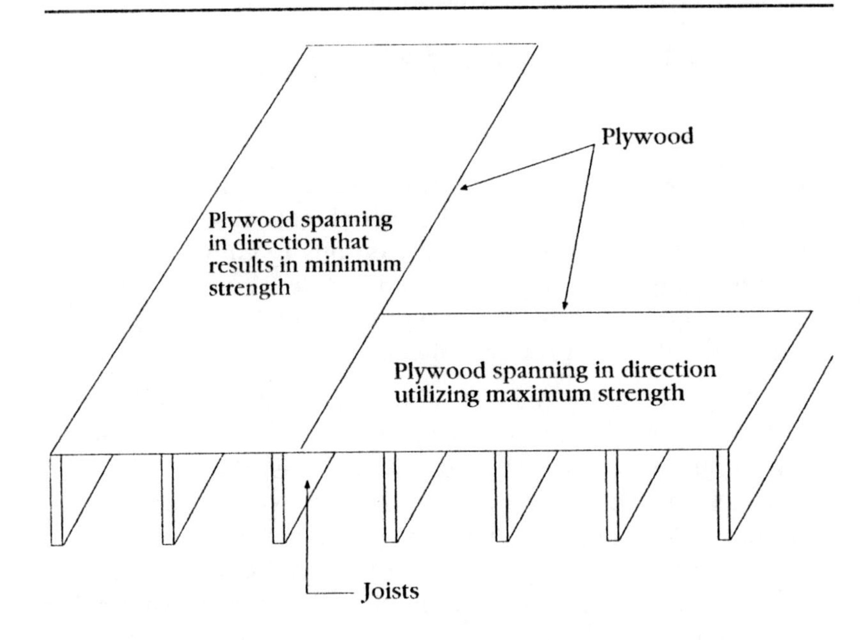

Figure 2-7—Plywood placed in two different orientations over joists

4. Exposure classification

For the grade of panel, the quality of the exposed sides will be noted by any combination of two of the A, B, C and D categories previously discussed. If the finish of the plywood on only one side is of interest, then only one letter would appear on the stamp. If the quality of either surface in unimportant for the application the stamp would have "Rated Sheathing" for a grade classification.

The allowable span length is designated on the stamp with two numbers, 24/16, for example. The first number denotes the allowable span length in inches for a piece of plywood used as roof sheathing. The second number denotes the allowable span length in inches for a piece of plywood used for subflooring. Note that both of these numbers assume that the long dimension of the panel is spanning across the supports.

Also noted on the APA stamp is the Exposure Classification: Exterior, Exposure 1, Exposure 2 or Interior, as well as the plywood thickness. Plywood thicknesses range from 1/4 in. to slightly over 1 in.

Advanced Discussion

It would appear that the technical aspects of plywood design would not be overly complicated. Surprisingly, this is not the case. The actual structural analysis and design is complicated and somewhat empirical. On the positive side, those involved in creating design specifications for plywood have been successful at making the selection process quite simple.

The following discussion will look at what is involved in the calculations used to determine the proper plywood for the loading condition it will be subjected to. As just mentioned, plywood analysis and design is complicated and is not easy to understand for those not trained in the engineering profession. For this reason, this discussion cannot fully cover all design and analysis possibilities but instead will try to present some of the aspects that need to be considered.

When designing plywood, as well as other structural members, the following must be considered:

1. Bending stresses
2. Deflections
3. Shears

The bending stresses can be produced by concentrated or by uniformly distributed loads. The discussion will only present the latter.

The bending stress is given by the following equation:

$$\text{Bending stress} = F_b = M/S$$

where M = Bending moment
S = Section modulus

The maximum bending moment for a simple span (see Figure 2-8A) is given by the following equation:

$$\text{Bending moment} = w\ell^2/8$$

where w = Uniform applied load
ℓ = Span length

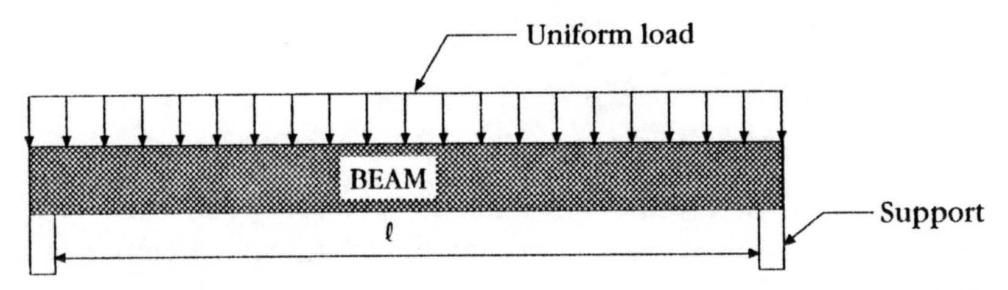

8a. Single-span beam

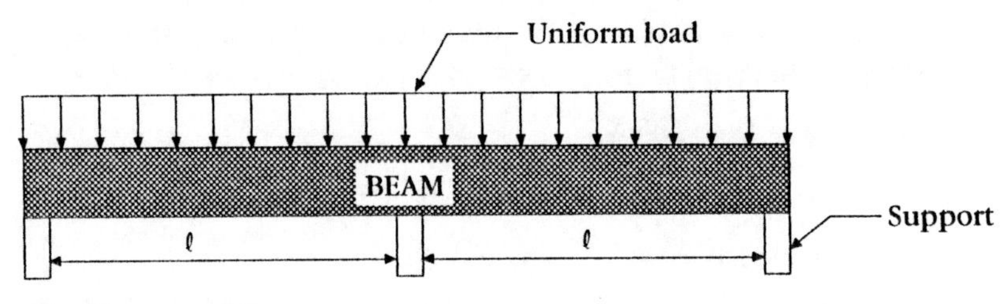

8b. Two-span beam

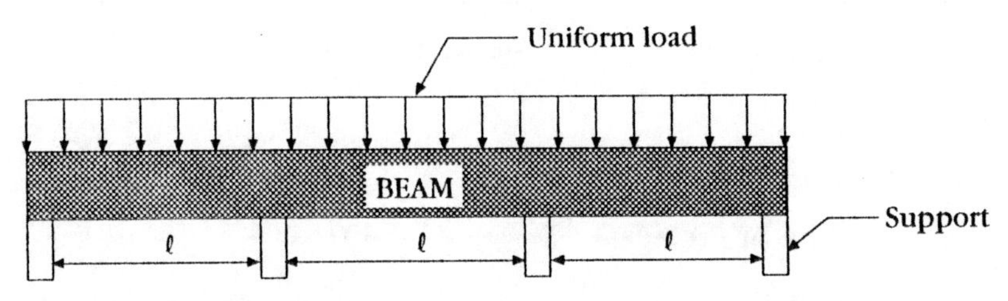

8c. Three-span beam

Figure 2-8–Single, two-span and three-span beams

Combining the two equations gives the following:

$$F_b = w\ell^2/8S$$

Rearranging this equation gives the following:

$$w = 8F_bS/\ell^2$$

If this procedure is used for a two-span condition (see Figure 2-8b) and a three-span condition (see Figure 2-8c), the following results are obtained.

$w = 8F_bS/\ell^2$ 1 span

$w = 8F_bS/\ell^2$ 2 span

$w = 10F_bS/\ell^2$ 3 span

Therefore, to determine the load the plywood can carry, the allowable bending stress, section modulus and span length must be known.

The value of the allowable bending stress for plywood used for flooring or roofing is 1650 psi. This value must be adjusted by the proper modification factors. Two of these adjustment factors are:

1. Duration of loads, DOL

Short-term loads	use 2.0
Wind loads	use 1.6
Snow loads	use 1.15
Normal	use 1.0
Dead Load	use 0.9

2. Panel Size, C_s

24 in. or greater	$C_s = 1.0$
20 in.	$C_s = 0.876$
16 in.	$C_s = 0.75$
12 in.	$C_s = 0.63$
8 in.	$C_s = 0.5$

The section modulus determination is typically easy to calculate. It would normally be:

$$S = bd^2/6$$

where b = width, use 12 in.
d = depth

For a 1-in.-thick piece of plywood, the section modulus per foot is:

$$S = 12(1)^2/6 = 2.0 \text{ in.}^3$$

However, this value of S is not used in plywood design. Rather, the value of S is adjusted downward to take into account that the plies are oriented in different directions, different species can be used for inner and outer plies and all plies cannot equally resist the loads. For this reason, the value of the section modulus for unsanded panels are:

Table 2-4—Section Modulus for Plywood

Thickness	Modified Section Modulus ($in.^3$)
5/16	0.112
3/8	0.152
15/32	0.213
1/2	0.213
9/32	0.379
5/8	0.379
23/32	0.496
3/4	0.496
7/8	0.678
1	0.859
1-1/8	1.047

Consider the following example:

Determine the maximum roof load that each of the plywood panels shown in Figure 2-9 can support in bending.

Use 7/8-in.-thick plywood

1. Three span, ℓ = 16 in.

   $S = 0.678 \text{ in.}^3$

   $F_b = 1650 \text{ lb/in.}^2$

   $$w = 10 \times 1650 \times 0.678/(16)^2$$
   $$= 43.7 \text{ lb/(in.)(ft)}$$
   $$= 524.4 \text{ lb/ft}^2$$

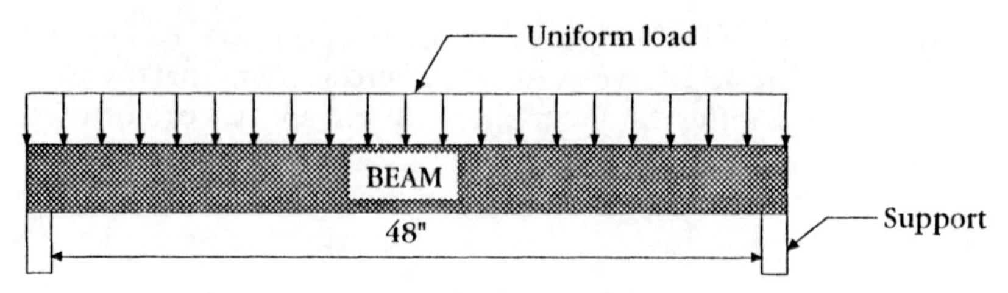

9a. Single-span beam

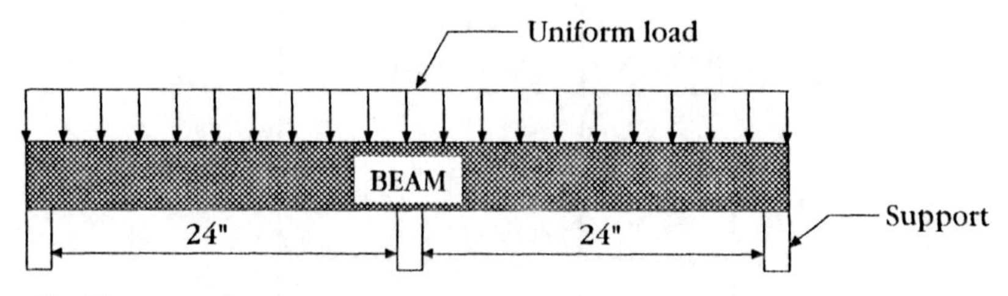

9b. Two-span beam

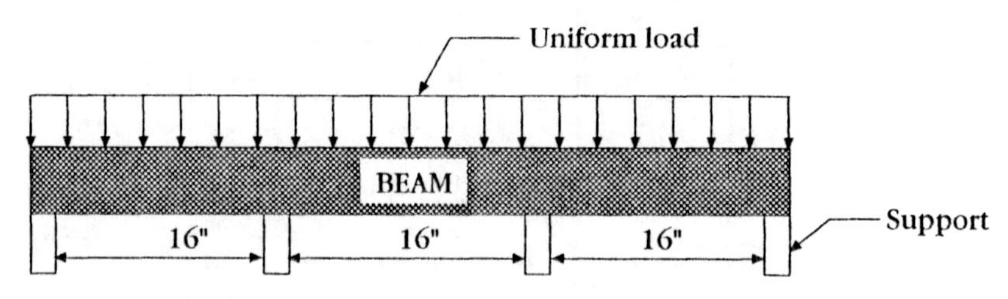

9c. Three-span beam

Figure 2-9—Loading conditions for examples

Assume the dead load is approximately 12 lb/ft^2

$$W_L = 524.4 - 12 = 512.4 \text{ lb/ft}^2$$

2. Two span, $\ell = 24$ in.
   $S = 0.678$ in.^{3}
   $F_b = 1650$ lb/in.^{2}

$$\begin{aligned} w &= 8 \times 1650 \times 0.678/(24)^2 \\ &= 15.5 \text{ lb/live ft}^2 \\ &= 186.5 \text{ lb/ft}^2 \end{aligned}$$

Assume the dead load is approximately 12 lb/ft^2

$$W_L = 186.5 - 12 = 174.5 \text{ lb/ft}^2$$

3. One span, $\ell = 48$ in.

$S = 0.678$ in.^{3}

$F_b = 1650$ lb/in.^{2}

$$w = 8 \times 1650 \times 0.678/(48)^2$$
$$= 3.9 \text{ lb/(in.)(ft)}$$
$$= 46.6 \text{ lb/ft}^2$$

Assume the dead load is approximately 12 lb/ft^2

$$W_L = 46.6 - 12 = 34.6 \text{ lb/ft}^2$$

Deflection of the plywood is also an important consideration. The maximum deflection for a simple span under a uniform load is given by:

$$\text{Maximum deflection} = \frac{5}{384}\,\frac{w\ell^4}{EI}$$

The maximum allowable deflection for roof live load is the span length divided by 240. The maximum allowable deflection for floor live load is the span length divided by 360.

Substituting the deflection equation and rearranging gives the following:

$$w = \frac{0.31EI}{\ell^3} \quad \ldots\ldots \quad \text{Roof live load}$$

Note that the term ℓ is the clear span length and the $\ell/240$ is the center-to-center spacing. The equation is simplified by using the center-to-center span length minus 1 in. Also

$$w = \frac{0.21EI}{\ell^3} \quad \ldots\ldots \quad \text{Floor live load}$$

For a two-span condition the maximum allowable distributed load is:

$$w = 0.77EI/\ell^3 \quad . \ . \ . \ . \quad \text{Roof live load}$$

$$w = 0.51EI/\ell^3 \quad . \ . \ . \ . \quad \text{Floor live load}$$

For a three-span condition the maximum allowable distributed load is:

$$w = 0.60EI/\ell^3 \quad . \ . \ . \ . \quad \text{Roof live load}$$

$$w = 0.40EI/\ell^3 \quad . \ . \ . \ . \quad \text{Floor live load}$$

In the discussion on bending stresses it was noted that the section modulus was reduced because of orientation of the plies, differing ply material, etc. For these same reasons, the typical calculation value of I is not used. The values of I used for plywood are shown in the table:

Table 2-5—Modified Moment of Inertia for Plywood

Thickness	Modified Moment of Inertia, I (in.^{4})
5/16	0.022
3/8	0.039
15/32	0.067
1/2	0.067
9/32	0.121
5/8	0.121
23/32	0.234
3/4	0.234
7/8	0.340
1	0.493
1-1/8	0.676

Consider the following example:

Determine the maximum roof live load that each of the plywood panels shown in Figure 2-9 can support based on deflection requirements. Use 7/8-in.-thick plywood.

1. Three-span, $\ell = 16 - 1 = 15$ in.

   $I = 0.340 \text{ in.}^4$

   $E = 1{,}800{,}000 \text{ lb/in.}^2$

   $$W_L = 0.6 \times 1{,}800{,}000 \times 0.34/(15)^3$$
   $$= 108.8 \text{ lb/(in.)(ft)}$$
   $$= 1305.6 \text{ lb/ft}^2$$

2. Two span, $\ell = 24 - 1 = 23$ in.

   $I = 0.340 \text{ in.}^4$

   $E = 1{,}800{,}000 \text{ lb/in.}^2$

   $$W_L = 0.77 \times 1{,}800{,}000 \times 0.34/(23)^3$$
   $$= 38.7 \text{ lb/(in.)(ft)}$$
   $$= 464.8 \text{ lb/ft}^2$$

3. One span, $\ell = 48 - 1 = 47$ in.

   $I = 0.340 \text{ in.}^4$

   $E = 1{,}800{,}000 \text{ lb/in.}^2$

   $$W_L = 0.31 \times 1{,}800{,}000 \times 0.34/(47)^3$$
   $$= 1.8 \text{ lb/(in.)(ft)}$$
   $$= 21.9 \text{ lb/ft}^2$$

Other items that would need to be considered would be concentrated loads and shear effects. These are considered beyond the scope of this book.

3

Steel

Introduction

Structural steel is used sparingly in residential construction projects, with the more predominate materials for structural members being wood and concrete. There are two possible reasons for the sparse use of structural steel. First, because of the high density of steel, the weight increase over other materials is significant. This extra weight can result in design difficulties, especially in regard to the soil pressure. Secondly, steel might not be cost-effective for a residential structure. In general, steel structures have greater material costs and lower construction labor costs. In contrast, concrete and wood tend to be the opposite. It would appear that the residential construction industry has found it more cost effective to utilize the materials that are cheaper and have greater associated labor costs.

Although it is not economical to use steel members for the majority of the framing, there are situations where its use is warranted. Steel beams are used to support wall loads and reduce the span length of the wood joists. Steel beams would be used if wood beams are impractical because of overstress or deflection.

Considerations

This chapter provides an overview of structural steel and its use in residential construction. The following topics are covered:

- What structural steel is used for in residential construction
- Structural shapes
- Material

Structural steel is any steel used for construction that is not embedded in concrete. If the steel is embedded in concrete, it is most likely concrete reinforcing steel. The structural steel members that are used for residential construction are:

Beams–Used beneath the first floor to support the loads of the upper floors and walls and in garage roofs. These beams are supported on the concrete basement walls and/or on steel columns.

Columns–Usually, the columns are hollow pipe and are used to support steel beams or other conditions where there are significant loads.

Lintels–Horizontal members placed above large openings (such as door or window openings in masonry) to provide support to the loads above the discontinuity.

Flitch plate–A steel plate that is attached to or sandwiched between wood beams. The steel and wood act compositely to provide a member with superior strength.

Steel shapes first became available in the early 1800s. The steel shapes that we are more accustomed to became available slightly over 100 years ago. When these steel shapes were first introduced, there was no standardization of the sizes for a given shape. In 1896, a steel association provided standardized dimension and nomenclature for given shapes. The shapes that became standardized were:

- S
- W
- HP
- Angle

- Structural pipe
- Structural tubing
- Channels

The S shape was the first standardized shape available. An S-shape member is shown in Figure 3-1. These members are actually I shaped but have been traditionally called *S shapes* denoting that they are standard sections. A sample designation of an S-shape member is:

S24 × 75

where S denotes that it is a standard section
24 denotes the depth of the steel member in inches.
75 is the weight in pounds per foot.

The W shape is the most common shape used in the construction industry (Figure 3-2). It accounts for more than half of all steel members in the construction industry.

The W designation is used because the member is called a *wide flange section*. Comparison of Figures 3-1 and 3-2 shows that the two shapes are very similar. The difference is that the S shape has a narrower flange and is sloped on the underside. A sample W-shape designation is:

W10 × 88

where W denotes it is a wide flange shape
10 denotes the depth of the steel member in inches.
88 is the weight in pounds per foot

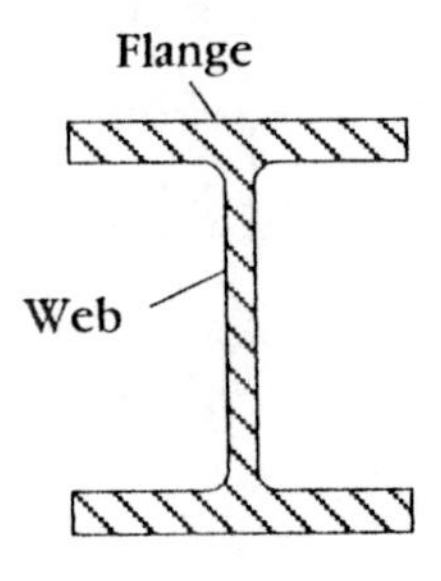

Figure 3-1—S-section steel beam

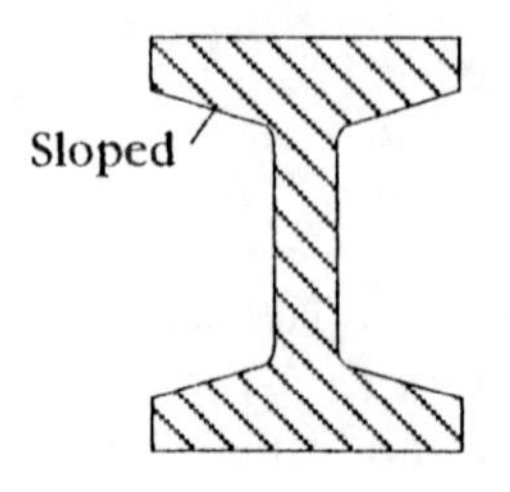

Figure 3-2—W-shaped steel beam

The HP shape is also similar in shape to the two previously discussed (Figure 3-3). This shape is used exclusively for foundation piles. Foundation piles, which can be wood, steel or concrete, are driven into the ground for the purpose of providing adequate foundation support for the structure. Pile drivers are used to hammer in piles to the proper depth. Because of this hammering these members must be resistant enough to take a significant pounding. This is why HP shapes, although similar in shape to W and S shapes, have thicker cross sections. A sample designation for HP shapes is:

HP14 × 89

where HP denotes it is an HP shape
14 denotes the depth of the steel member in inches.
89 is the weight in pounds per foot

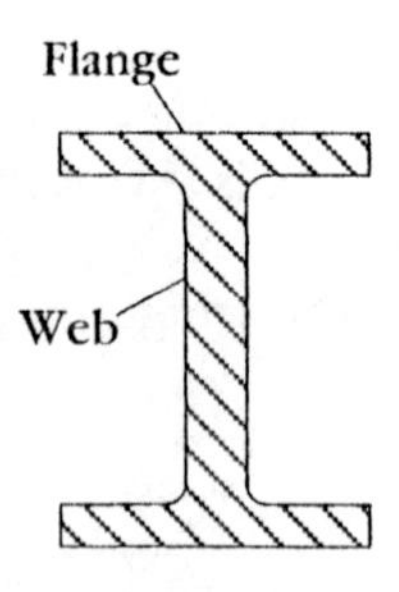

Figure 3-3—HP-section steel beam

The angle shape (Figure 3-4) is one of the most versatile of the steel shapes. It can be used in practically all types of construction. A sample designation is:

L5 × 3-1/2 × 3/4

where L denotes that it is an angle
5 is the length of the longer leg in inches
3-1/2 is the length of the shorter leg in inches.
3/4 is the thickness of the member

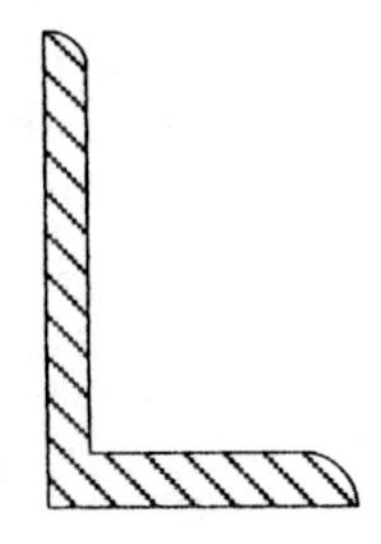

Figure 3-4—Angle steel beam

Structural pipe (Figure 3-5) is useful as a column since it is relatively easy to install. Pipe diameters come as small as 0.5 in. and as large as 12 in. in diameter. These columns are hollow with thicknesses from 1/8 in. to 3/8 in. These columns can be purchased as standard weight, extra strong or double-extra strong. An extra-strong pipe has a wall thickness of 35 to 50% greater than the standard weight pipe. A double-extra strong pipe has a thickness of almost three times that of a standard pipe. Steel pipes are designated as follows:

3-1/2 in. standard weight pipe

where 3-1/2 in. denotes the inside diameter of the pipe. Therefore, the outside diameter is larger.

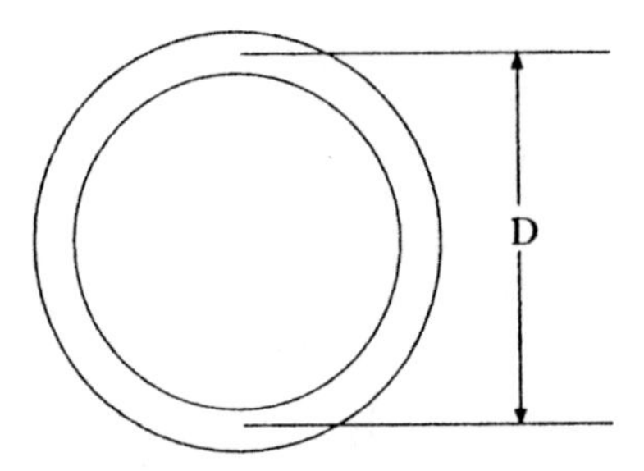

Figure 3-5—Structural steel pipe

Structural tubing comes in square or rectangular shapes (Figure 3-6). Structural tubing is often used in light construction. Steel tubing is designated as follow:

4 × 3 × 0.25-Structural Tubing

where 4 is the long-side dimension
3 is the short-side dimension
0.25 is the wall thickness of the tube

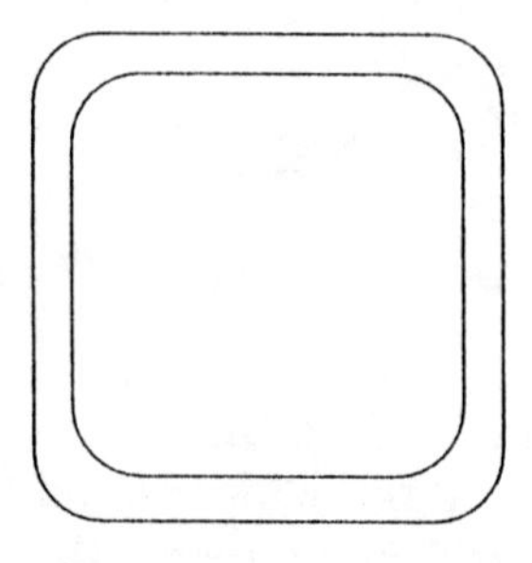

Figure 3-6—Structural steel tubing

The last structural shape to be discussed is the channel (see Figure 3-7). Channels serve a useful purpose since they have flanges which provide adequate strength, but have flanges only on one side. The one-sided flange allows for the easier concealment and facilitates abutting up against other structural elements. A sample channel member designation is:

C9 × 15

where C denotes that the member is a channel
9 is the depth of the channel
15 is the weight per foot in pounds

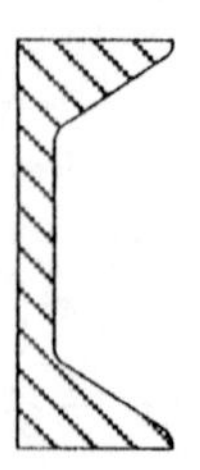

Figure 3-7—Channel steel section

This concludes the discussion of the various structural steel shapes. The following is a discussion on how each of these shapes relates to residential construction.

S-shape–Not used in residential construction.

W-shape–Accounts for the majority of the steel beams used in residential construction. Beams are typically 8, 10 or 12 in. deep.

HP-shape–Not used in residential construction.

Angles–Used as lintels to support masonry loads over discontinuities. Usually utilize angles with a 3-1/2 in. horizontal leg.

Structural pipe–Used as columns to support beams or other heavy loads. Outside diameter should not exceed 3-1/2 inches to allow pipes to be concealed in a conventional stud wall.

Structural tubing–Not used in residential construction.

Channels–Seldom used in residential construction. May be used to provide greater load-carrying capacity to headers over large openings.

Steel is classified by the use of an "A" designation. This A designation denotes the ASTM (American Society of Testing Materials) standard which the steel must satisfy.

Prior to the 1960s there was only one grade of structural steel for construction. Presently there are seven types of structural steels. These types can be placed into four categories based on their chemical composition and corrosion capabilities.

The first category are steels that rely on the carbon content for their strength. This category represents the least expensive and not surprisingly the most popular structural steel. The two steels in this category are A36 and A529.

The second category is high-strength steels. Various alloys are added to the material that acts in conjunction with the carbon to provide a high-strength steel. A441 and A572 are the steels that compose this category.

The third category is actually the same as the second category, except that the chemical make-up provides a corrosion-resistant material. A242 and A588 make up this category of steel.

The fourth and final category is steel that has both carbon and alloying elements and is quenched and tempered. The quenching and tempering provide an extremely strong steel. A514 is the designation for steel in this category.

Figure 3-8 provides a comparison of the yield strength of the various steels.

As previously mentioned, the most popular steel is A36. Analysis of Figure 3-8 shows that A36 has the lowest yield strength. A 36 is the exclusive steel used in residential construction, but obviously any other steel would work as well. A572 and A588 are high-strength steels that are popular, along with A36, for bridges and high-rise buildings. The other steels see limited action in construction.

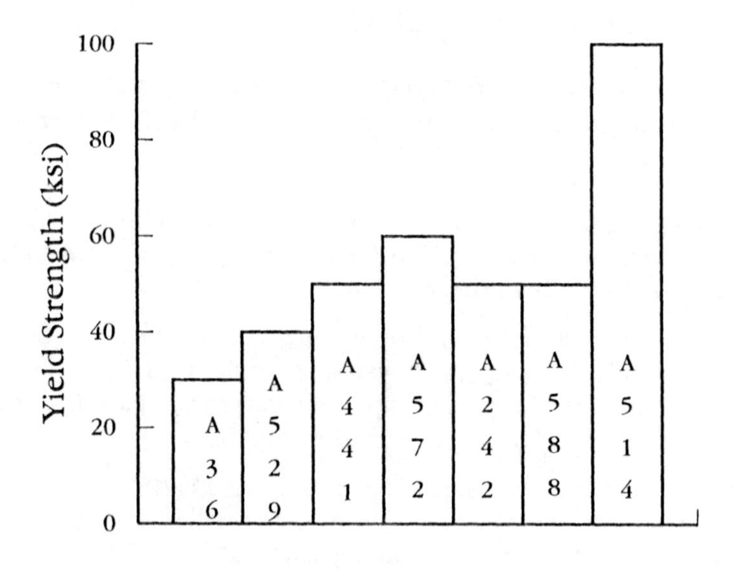

Figure 3-8–Steel strengths

Advanced Discussion

The individual steel members in a structure must be properly tied together to obtain a contiguous, rigid structure. The steel connections required for residential construction are typically limited to the beam-to-pipe column connection and the pipe column-to-foundation connection. These connections are relatively unsophisticated and require minimal thought and planning.

Situations do arise in residential construction where a more sophisticated connection is needed, but this is not the norm. It would be more likely that changes required for a remodeling project would require a complicated connection.

Two methods are available for connecting steel members: bolting and welding. An overview of these methods is presented in this discussion.

Bolts, similar to steel members, have ASTM designations to distinguish bolts with different properties. Bolts used for structures are divided into two categories:

1. Unfinished bolts
2. High-Strength bolts

Unfinished bolts are designated as ASTM A307. These bolts are often called *common bolts* and are the least expensive. These bolts, which are used in residential and light construction, are easily spotted since they often have square heads and nuts.

High strength is needed when the connection is a significant part of the integrity of the structure. High-strength bolts are either ASTM A325 or ASTM A490. A325 bolts, similar to normal-strength steels, utilize carbon to gain their high strength. A490 bolts, similar to higher-strength steels, use alloys to gain their extra strength. A490 bolts are approximately 25 percent stronger than A325 bolts for smaller diameters and 50 percent stronger for larger diameters. A high-strength bolt can usually be distinguished from an unfinished bolt since the former has a hexagonal head rather than square head.

The purpose of bolting is to tie the steel pieces together. The tying force is provided by either a friction-type or bearing-type connection. From an appearance viewpoint, both types of connection look identical. A bearing-type connection is designed such that the bolt nor the edge of the hole it bears against is overstressed. This type of bolted connection is not suitable where there are load reversals or vibrations. Under these conditions, a friction type connection is used. The bolts in a friction-type connection are tightened to a degree that they will not slip and subsequently do not bear against the edge of the hole.

Regardless of which type of connection is used, the bolts must be sufficiently tightened to provide the necessary clamping force. Three methods are available to ensure that the bolts are adequately tightened. These methods are:

- Direct tension indicator–The nut and bolt utilize a washer that is not flat. As the bolt is tightened the nut flattens the washer. When the washer is flush with the steel member, the bolt has been properly tightened.
- Turn-of-the-nut method–The nut is tightened as much as possible by hand. The workman will then turn the nut a predetermined number of turns with a tool to obtain the proper tightening amount.
- Calibrated wrench–A tool is used that will stop tightening the bolt after it has reached a preset torque level.

The other method used to join steel members is welding. Welding is the joining of two pieces of metal by heating them and connecting them with a filler metal. Welding is typically used in residential construction to connect plates to the base and top of the pipe column.

The most common types of weld joints are shown in Figure 3-9.

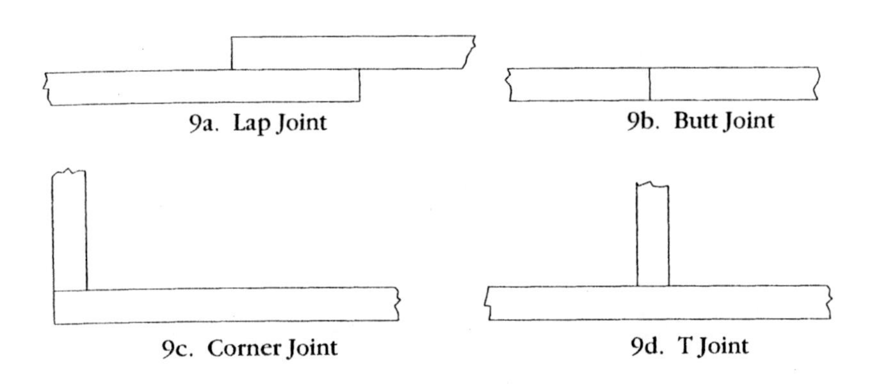

Figure 3-9—Types of welded joints

These joints are accomplished using the welds shown in Figure 3-10.

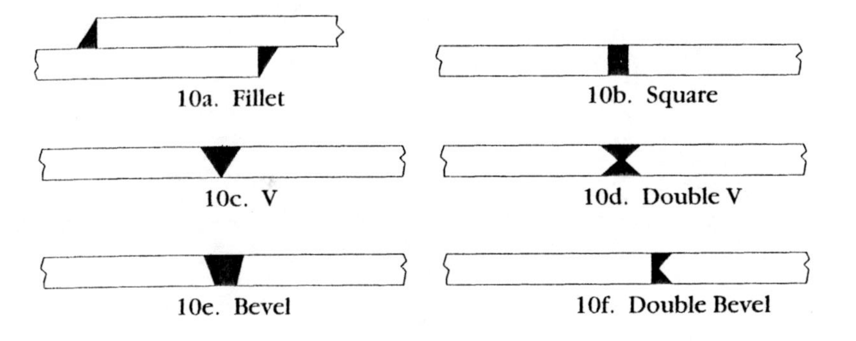

Figure 3-10—Types of welds

A proper weld is a function of the following:

1. Use of proper welding equipment and electrodes.
2. Sufficient fusion of the base material and the weld metal.
3. Avoiding cracking or porous welds.

4

Concrete

Introduction

Concrete is an excellent material for many types of construction projects. Concrete is durable and strong and can be formed into unlimited shapes and sizes. In residential construction, concrete can be used for:

- Foundations
- Sidewalks
- Stairs
- Slabs
- Driveways

Concrete is relatively cheap as a material, but will include significant labor costs in its placement. However, the end product is durable and has a long service life, which will offset any large initial costs.

Concrete is composed of a mixture of the following materials:

- Cement
- Water
- Sand
- Stone

Cement is a fine powder that provides the bond so that the materials stick together to form a monolithic substance. Cement is available in the following types:

Type 1–Common cement used in the construction of buildings, bridges, sidewalks and pavements.

Type 2–Cement used in structures that will be located in water.

Type 3–Cement that gains strength rapidly.

Type 4–Cement used in massive structures (i.e., dams).

Type 5–Cement used in structures where sulfate exposure is likely.

Type 1 is the most common type of cement used in residential construction.

Water reacts chemically with the cement to make a paste that will solidify the mass. Any water that is suitable for use as drinking water is suitable for use in concrete. The water used in the mix is of concern since it may contain undesirable chemicals that may have detrimental reactions with other materials in the mixture.

It is possible to make concrete by using only cement and water, although it would be extremely expensive. For economic reasons a filler is used to reduce the amount of cement needed. The filler used is sand and stone since both fine and large particles are needed to prevent voids. This filler material accounts for more that two-thirds of the volume of the concrete.

Concrete members can be either cast-in-place or precast. For cast-in-place members the concrete is delivered by truck and the member is placed on site. Precast members are constructed off-site under controlled conditions and after curing are transported to the site and lifted into place. The great majority of concrete in residential concrete is cast in place.

The greatest distinction in concrete is whether it is reinforced or unreinforced. Concrete is strong in compression. A load that tends to squeeze a member is a compressive force. A load that tends to stretch a member is a tensile force.

Concrete is very poor in resisting tensile forces. Figure 4-1 shows a concrete member subject to both tension and compression. Cracking will occur in tension for a load of one-tenth of what it takes to cause cracking in compression.

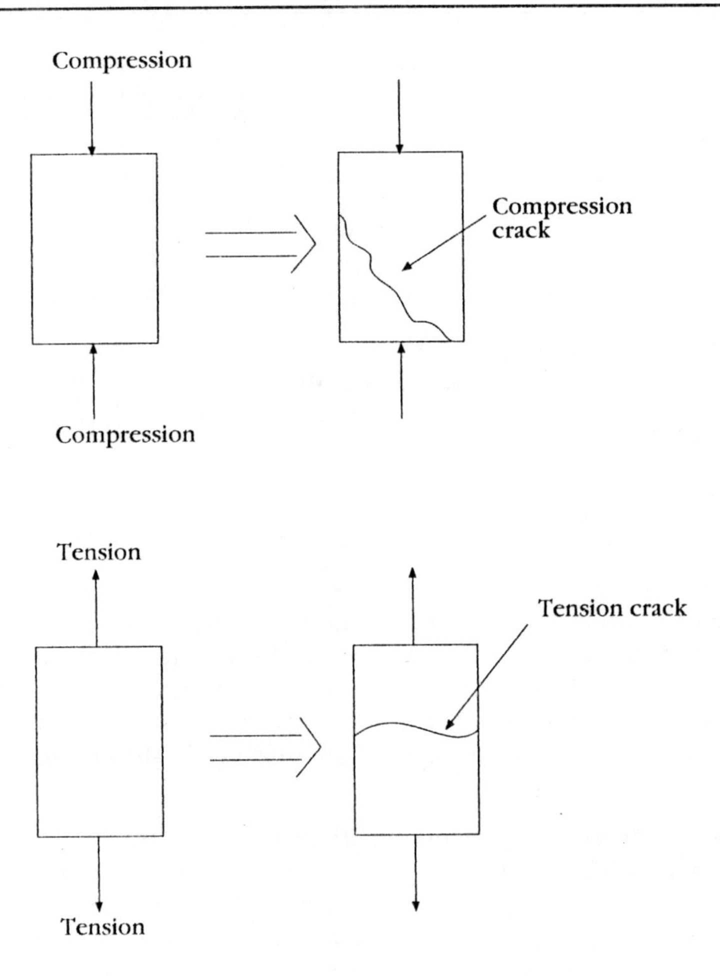

Figure 4-1—Member loaded in compression and tension

Compression and tension stresses do not always occur separately. When a beam bends, the top part is in compression; that is, it will shorten. The bottom part is in tension and is elongated. The shortening and elongating is depicted in Figure 4-2. Under these conditions the beam will crack on the bottom at a load far greater than what will cause cracking at the top. Therefore, if the load is kept at a value that cracking is prevented, the maximum allowable load is controlled by tensile forces. Allowing the tensile stresses to control the magnitude of the load is inefficient since most of the member is highly understressed.

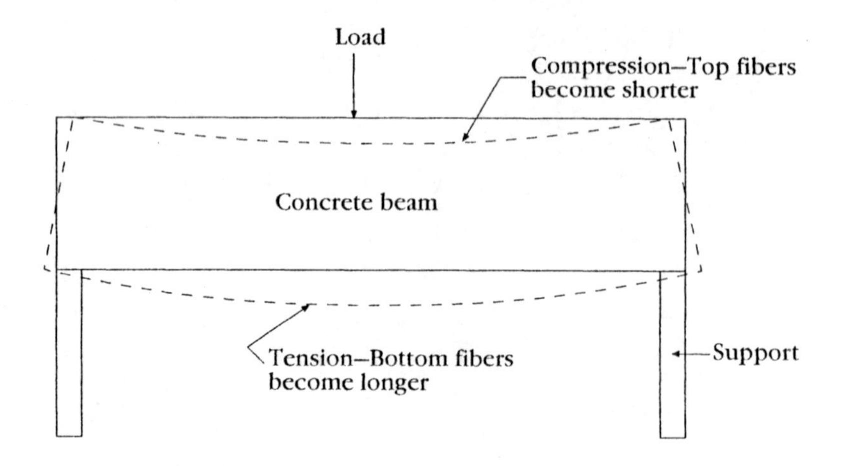

Figure 4-2–Compression and tension in loaded beam

This shortcoming of concrete can be overcome by using steel reinforcing. Steel reinforcing bars can be used to supplement the tensile strength of the concrete. Figure 4-3 shows a reinforced and an unreinforced beam. The reinforced beam uses steel reinforcing to provide additional strength. Concrete is typically unreinforced in residential construction.

One of concrete's strongest qualities is that it is durable. Concrete sidewalks, driveways and slabs must be able to withstand the effects of freezing and thawing cycles. The ability to resist the effects of weather is a function of the composition of the concrete. An approximate correlation between the materials and durability can be made as follows:

Water content–Increased water, lower durability.

Cement content–Increased cement, greater durability.

Stone and sand–No correlation.

Air entrainment–Improves durability

Air entrainment is the application of small air bubbles into the concrete mix. These air bubbles improve the performance of the concrete from a durability standpoint. When significant moisture is present, water will penetrate into the small natural voids in the slab. When this water freezes, it expands and induces stresses that might cause the top of the slab to break apart. If air entrainment

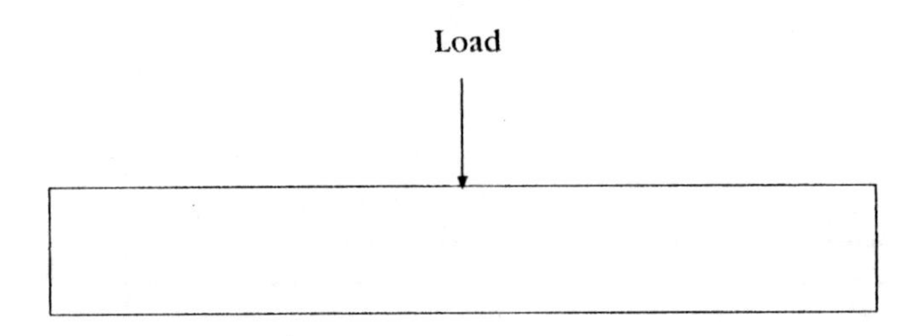

3a. Load on unreinforced beam

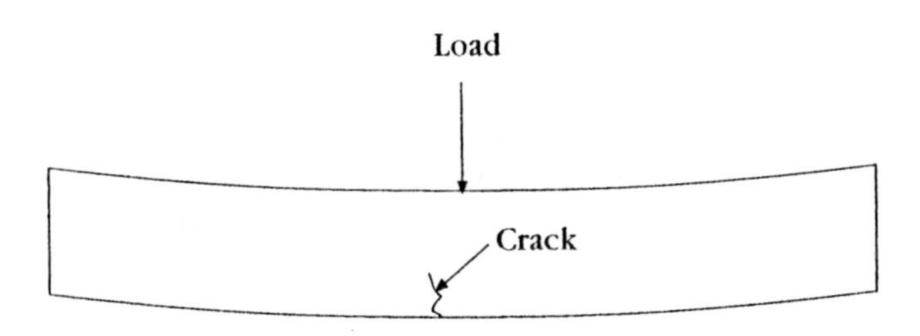

3b. Initiation of crack in unreinforced beam

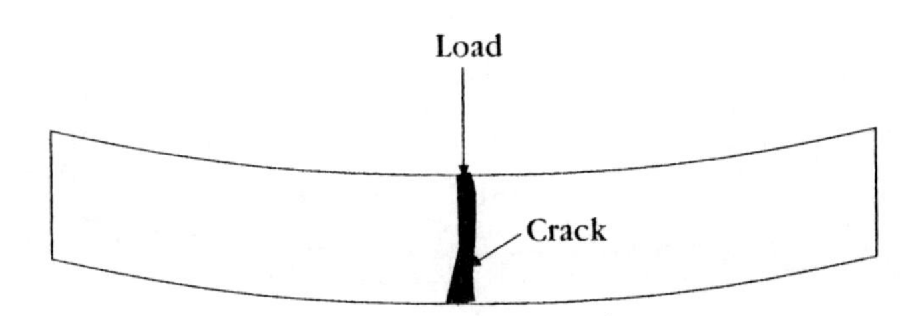

3c. Crack through entire depth of unreinforced beam

Figure 4-3–Cracking in unreinforced and reinforced concrete beams

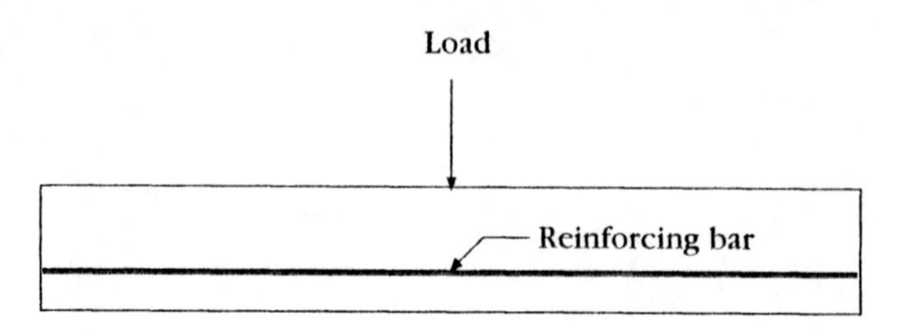

3d. Load on reinforced beam

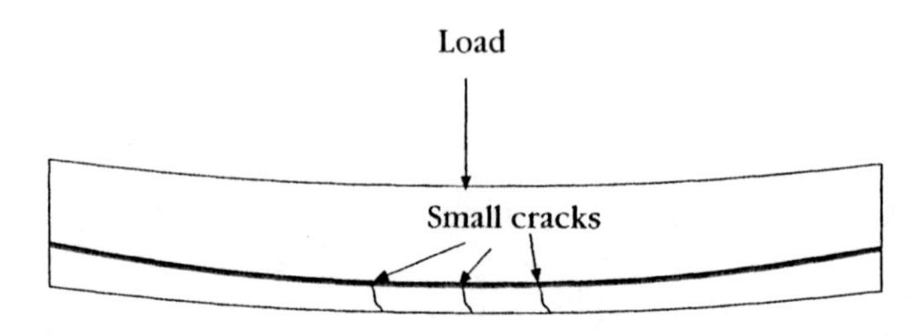

3e. Initiation of small cracks in reinforced beam

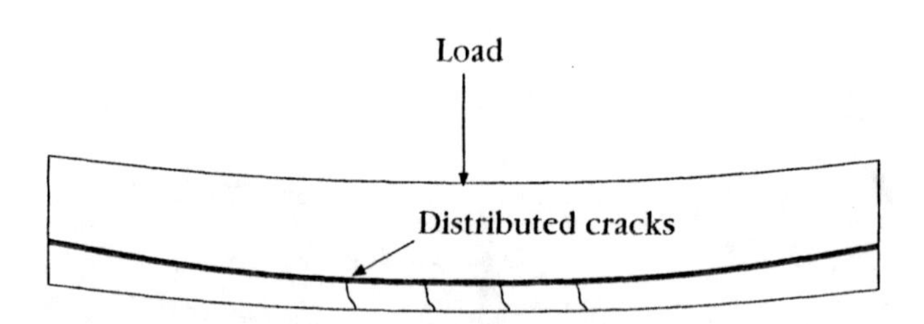

3f. Cracks under significant loads

Figure 4-3—Cracking in reinforced and unreinforced concrete beams (cont.)

is used, there are extra voids in the concrete. The freezing water can expand into these extra voids rather than pressing against the surface of the concrete.

Concrete must flow freely into place when a member is being constructed. The stiffer the mix, the more difficult it is to place the concrete. However, if additional water is added to the mix, the weaker the concrete, and the more prone it will be to freezing and thawing deterioration. Construction crews often add more water to the concrete mix to make it more workable. This water addition can produce a concrete that has less strength than required. If the ratio of the weight of water to the weight of concrete doubles, the strength of the concrete is approximately cut in half.

The physical characteristics of concrete that are of interest are:

- Compression strength
- Tensile strength
- Modulus of elasticity

Compressive strength is a measure of ability of concrete to resist loads that crush the concrete. The compressive strength is the most common parameter used to specify the concrete. Compressive strengths range from as low as 2000 psi to as high as 20,000 psi. It is unusual to have a concrete with a strength over 6000 psi specified. Concrete strengths of 6000 psi to 10,000 psi are used in the columns and walls of high-rise construction. Concrete strengths in excess of 10,000 psi are very rare. In residential construction, compressive strengths used are 2500 to 3500 psi.

It is important to note that the compression strength of the concrete that is specified corresponds to the strength at 28 days. As concrete ages, it gets stronger. The strength at 29 days will be greater than the strength at 28 days. Consider the following example, given that the compressive strength of the concrete is requested to be 4000 psi. The following are the test results for the concrete at various ages.

Time	Strength
Day 0	0
Day 7	3200
Day 14	3600

Day 21	3800
Day 28	3900
Day 35	4000
Day 42	4050
Day 108	4200

Although the concrete has a strength of 4000 psi at day 35, it does not qualify as 4000-psi concrete. The strength of concrete is measured at 28 days. Therefore, this concrete is considered to have a strength of 3900 psi.

Tensile strength is a measurement of the resistance of the concrete to forces that tend to stretch it. The tensile strength can be roughly estimated as one-tenth of the compressive strength.

The deflection of a member is inversely proportional to the modulus of elasticity. If the modulus of elasticity is doubled, the deflection will be cut in half. The modulus of elasticity can be estimated as:

$$\text{Modulus of elasticity} = 57{,}000 \sqrt{\text{Conc. strength}}$$

Review of this equation reveals that a higher-strength concrete will have a corresponding higher modulus of elasticity.

Some useful tests and measurements of concrete properties include:

1. Slump test
2. Air content test
3. Compression test
4. Petrography

The slump test is a measure of the stiffness of the concrete mix. The test is started by filling an upside-down cone which has neither a top or a bottom. The upside-down cone is filled while it is held snugly against the ground. After the cone is filled to the top, it is pulled upward, allowing the concrete to flow laterally. The distance that the top of the cone-shaped concrete moves downward from its original position is called the *slump*. A concrete mix with a very high slump usually has a higher water content than a mix with a low slump. A concrete with a low slump will typically provide a more durable concrete but is more difficult to place.

The air content is performed to determine the amount of air voids in the concrete mix. The amount of air voids might be important, depending on the environment the concrete will be subjected to. One of the most common methods of measuring voids is by a pressure system. For this method the concrete is placed in an air-tight container. The measured air pressure gives an indication of the air voids and the corresponding expected performance.

The compression test is an indication of the overall strength of the concrete. A compression test is performed by placing a concrete cylinder in a machine that will crush it. The load is applied slowly to the cylinder until it splits. The cylinders are about 1.5 ft high and 0.5 ft in diameter.

Petrography is one of the most interesting aspects of testing of concrete. The term *petrography* is derived from the word *petro*, meaning *rock*. Those who practice the art of petrography, are called *petrographers* and typically have a background in geology. Petrographers take a sample of hardened concrete and grind a small piece until it has a very small thickness which doesn't appear to be thicker that this piece of paper. The petrographer will shine a light through this thin piece of concrete. From experience the petrographer can approximate how much water was used to make the concrete, the amount of voids and how much cement reacted with the water.

The ratio of the materials that form the concrete (water, cement, stone and sand) will determine how the finished concrete will perform. Another important aspect that will determine performance is the preparation of the location where the concrete will be placed. This mainly applies to slabs that are built on grade. A slab on grade must remain crack free so that it will not be unsightly, be a tripping hazard or allow water to permeate up through the slab. The slab must be placed on a prepared base which includes the following:

1. A good draining material.
2. Use a material that will not compress under the weight of the slab.
3. Compact the material so that it will not compress nonuniformly.

For walls, base preparation is not the important issue, proper formwork construction is. The formwork must have strong enough ties to prevent the formwork from spreading outward and collapsing. The plywood must be strong enough that it will not bulge outward between the ties.

The inside of the forms must be coated with a releasing agent to allow for the removal of the forms after the concrete has cured. Failure to use a releasing agent might cause the plywood to be permanently bonded to the concrete.

Although concrete is very durable and has a pleasant appearance, it can become quite undesirable if it has defects. Some of the more common defects and distresses in concrete include the following:

- Scaling of the top surface
- Cracking from shrinkage
- Popouts

Scaling of the top surface of the concrete is usually the result of substandard finishing by the tradesman. Excessive troweling of the top surface will cause the water to migrate to the surface. This will result in a larger water-to-cement ratio near the top. A large water-to-cement ratio will produce a concrete with a weaker surface that will fall apart when subjected to freezing and thawing cycles.

Cracking from shrinkage occurs generally from the slab being restrained during the drying process. As the water evaporates from the concrete, it will reduce in volume. If the slab is restrained, it will be prevented from shrinkage and will crack. This occurs for slabs on grade since water easily evaporates at the top surface while the bottom of the slab dries slower and is restrained from movement by the frictional force of the base material.

The final topic of defects is popouts. Popouts are random pits on the surface of the concrete. Popouts are usually caused by using expansive aggregates. Expansive aggregates are stones that expand when they absorb moisture. Stones that expand near the surface can generate enough force to break away the upper surface of the concrete. Popouts can be avoided by the concrete supplier not using any expansive aggregates in the concrete mix.

The discussion presented here will focus on three aspects of concrete construction:

1. Reinforcing bars
2. Prestressed concrete
3. Concrete additives

Advanced Discussion

As mentioned in the previous section, reinforcing bars are used sparingly in residential construction. The other two items listed, prestressed concrete and concrete additives are rarely used in residential construction. However, they are presented since they are very common to the rest of the construction industry and have a potentially bright future if introduced into the residential construction market.

Reinforcing bars are used in concrete to overcome the failure of concrete to perform well in tension. The necessary amount of steel reinforcing is dependent on the force exerted on the concrete member. Reinforcing bars are manufactured in the following standard sizes:

Table 4-1—Reinforcing Bar Sizes

Bar Size	Diameter (in.)	Area ($in.^2$)
3	0.375	0.11
4	0.500	0.20
5	0.625	0.31
6	0.750	0.44
7	0.875	0.60
8	1.000	0.79
9	1.128	1.00
10	1.270	1.27
11	1.410	1.56
14	1.69	2.25

Reinforcing bars for residential construction are typically no. 4, no. 5 or no. 6 bars.

Besides size, reinforcing bars are also specified by material strength. The parameter of material strength used is yield strength, which is the strength of the steel before it will be unable to carry any additional load without having uncontrollable deformation. Steel with yield strengths of 40,000 and 60,000 psi are the predominate steels used for reinforcing bars. These are called *Grade 40* and *Grade 60* bars, respectively. The majority of all reinforcing bars are composed of Grade 60 steel.

For the reinforcing bars to properly perform their function, they must act integral with the concrete. This is accomplished by chemical and mechanical bonding of the reinforcing bars to the concrete. Chemical bonding comes from the gluing effect of the cement attaching to the reinforcing bars. The mechanical bonding is achieved by providing bumps on the steel reinforcing bars. These bumps, called *lugs*, will bear against the adjacent concrete and prevent the reinforcing bars from movement. Lugs are located every few inches along the length of the reinforcing bar.

In cases where steel reinforcing bars can be subjected to a corrosive environment, the bars can be coated with a protective coating. This protective coating is usually an epoxy and will protect the steel reinforcing bars from chloride attack.

An innovative design concept that has become increasingly popular in just about every type of construction market, except residential, is prestressing. The theory behind prestressing is to introduce forces in the concrete member before it is placed into service. These forces will produce stresses that are opposite of the stresses that will occur from actual loading conditions. For example, a beam used in a bridge will deflect downward when loaded with trucks. Prestressing could be used in this case by introducing forces, before the beam is put into service, that will cause the beam to bend upward so that it will offset some of the deflection caused by trucks and other loads.

Figure 4-4 gives an example of a prestressed beam. Note that the great advantage of a prestressed beam is that by adding compression forces the possibility of cracking is reduced.

The process of prestressing is performed by taking steel strands and pulling them tight with a jacking force. The beam will tend to

4a. Unloaded conventional beam

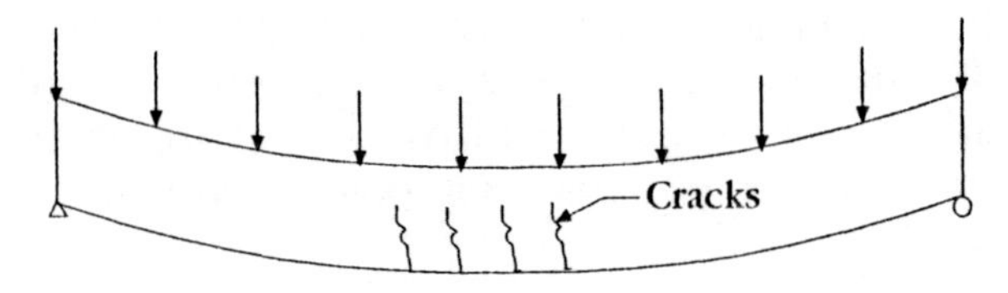

4b. Loaded conventional beam

4c. Unloaded beam (same as 1a.)

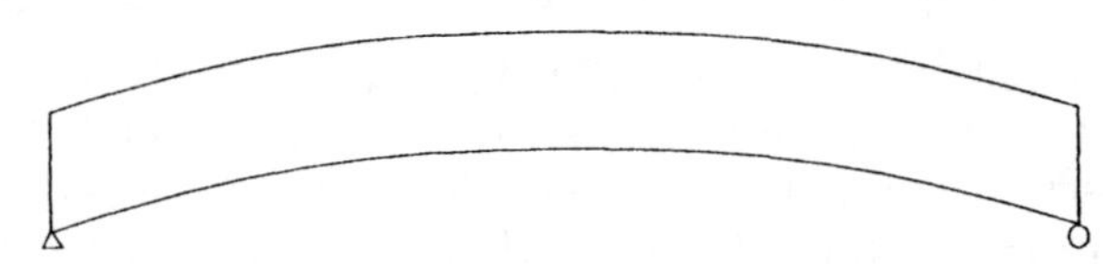

4d. Unloaded beam deflected upward from prestressing

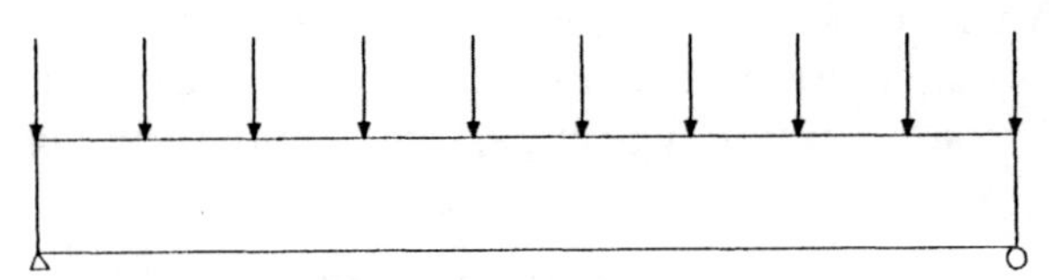

4e. Beam after loads are added

Figure 4-4—Prestressed compared to conventionally reinforced beams

bow upward. The strands can be placed in the formwork prior to the concrete being placed. After the concrete is placed and sets, the jacks are removed and the force in the strands are transferred to the concrete. When the strands are stressed prior to the placement of the concrete, the prestressing is called *pretensioned*. The steps involved in pretensioning are shown in Figure 4-5.

An alternate method is to place the strands in the formwork and place the concrete prior to stretching the strands. After the concrete hardens, the strands are pulled thereby introducing the needed stresses. This method of prestressing is called *post-tensioning*. The steps involved in post-tensioning are shown in Figure 4-6.

Additives can be placed in the concrete to provide a mix that has more desirable properties for installation and for its end use. Some of the more common concrete additives, also called *admixtures*, are:

- Setting accelerators
- Concrete coloring
- Superplasticizers

Setting accelerators help the concrete achieve high-strength at an early age. Setting accelerators are helpful for fast track projects where the concrete must dry and have sufficient strength for the next phase of construction to begin. For example, the concrete foundation walls must be allowed to dry for several days before the forms can be taken off and wood framing can begin. If a setting accelerator were used, the forms could be removed earlier and wood framing could begin very shortly. One would have to weigh the decrease in cost from saving time off the construction schedule with the costs of the setting accelerator. In some cases the use of setting accelerators is a must. In cases where repairs are needed in high-volume traffic highways, there is not time to have normal setting periods.

The cost associated with the use of a setting accelerator is not the only concern to the contractor. Setting accelerators are calcium chloride based. Calcium chloride is corrosive to any embedded steel, including steel reinforcing bars.

Additives used to color concrete have become increasingly popular over the last decade. Coloring concrete in conjunctions with

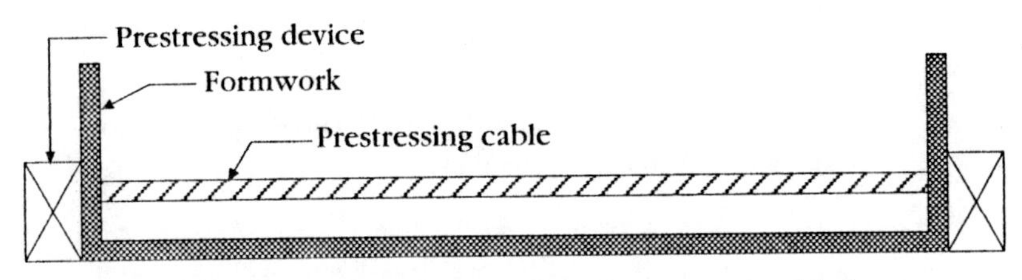

5a. Steel prestressing cable stretched before concrete is added

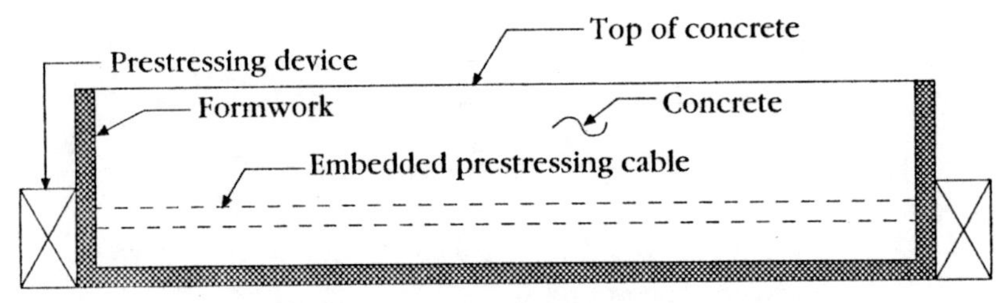

5b. Concrete placed in formwork with prestressing cable

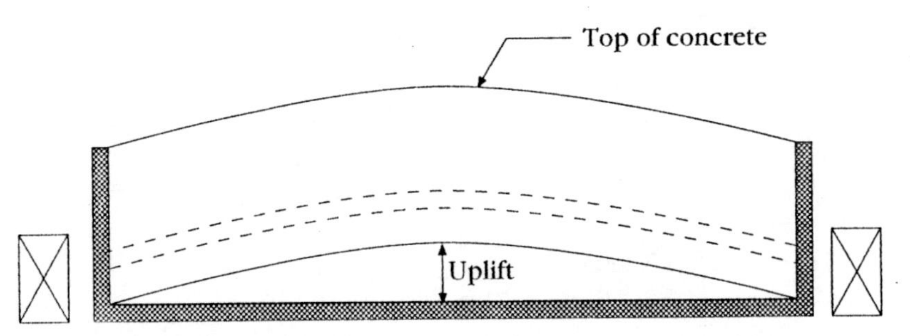

5c. After concrete has set, prestressing devices are removed and beam cambers upward

Figure 4-5—Pretensioned concrete beam

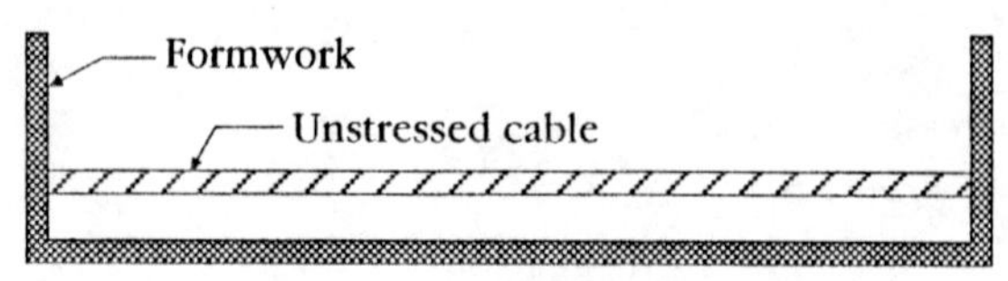

6a. Unstressed cable placed in forms

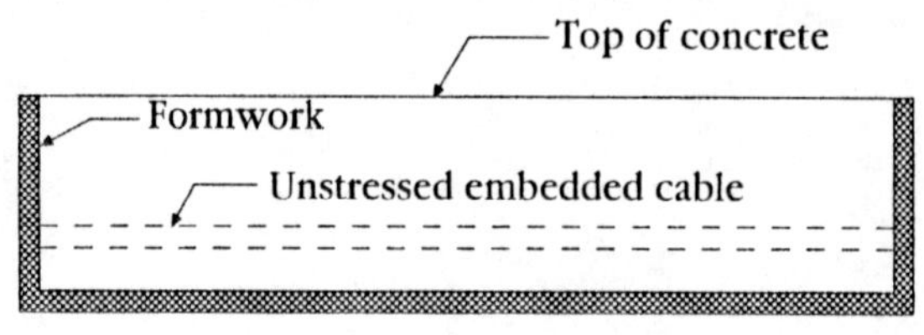

6b. Concrete added to unstressed cable

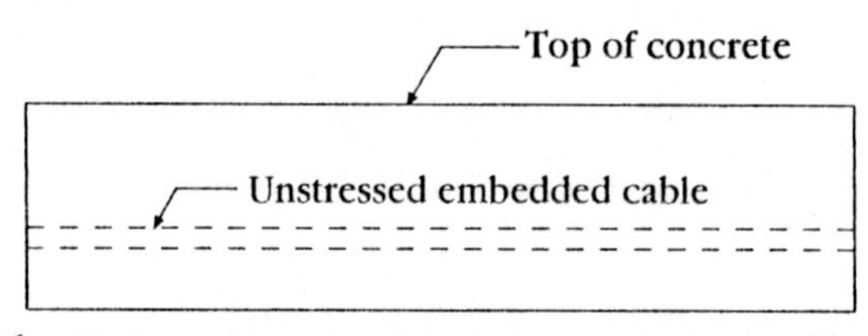

6c. Formwork removed

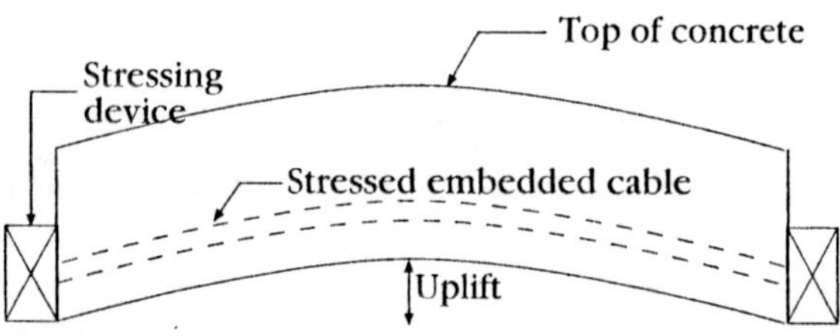

6d. Beams deflects upward when cable is stressed

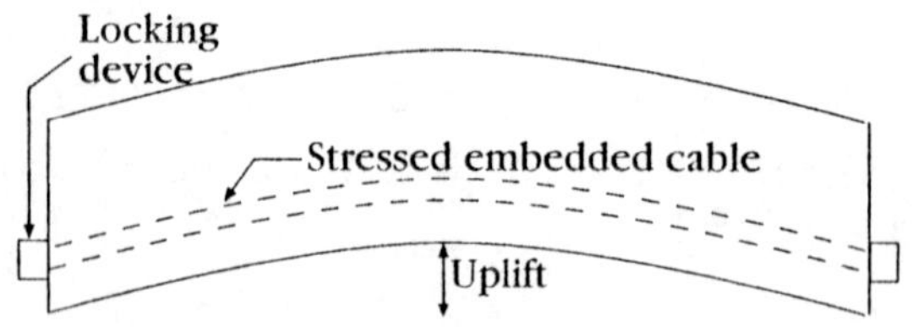

6e. Stressing devices removed and locking device used to hold concrete tight

Figure 4-6—Post-tensioned concrete beam

pattern stamping produces driveways and sidewalks that are quite impressive. Coloring of concrete is accomplished by adding different oxides to the mixture.

The more water that is added to the concrete mixture, the weaker the strength of the concrete. On the other hand, the more water that is added, the easier it is to place the concrete and it will tend to have better flow characteristics. One method of getting a flowable concrete without having to use water is to use superplasticizers. Superplasticizers can produce high-slump concrete without the use of any additional water.

5

Masonry

Introduction

The use of masonry in construction had its origin many centuries ago when mixtures of straw and moist clay were sun dried into building blocks. Masonry is presently manufactured in a much more sophisticated manner and has more sophisticated uses. It has developed a niche in construction that transcends its original use as a building frame. Masonry is used for fences, paving, basement walls, high-rise building infill panels and veneer. The discussion in this chapter is limited to masonry as a veneer and for walls.

Brick is considered veneer rather than a structural element when it is used as a closure element. Wood joists, studs and plywood are designed to resist the forces applied to the building and transfer the reactions to the foundation. The building would remain structurally sound with the use of just these framing elements. However, to make the building practical it must be enclosed to shield the occupants from the environment. This is accomplished by covering the frame with non-structural elements. Nonstructural elements include wood siding, vinyl siding and brick. If brick is used to perform this enclosure function, it does not have a structural function and is referred to as *veneer*.

The brick veneer carries the load of its own weight directly to the foundation. It does not rely on the wood frame to carry any of its weight. But since the brick wall is tall with minimum thickness, it has little strength against lateral forces. If the masonry is left free standing, it would easily topple over. Lateral stability for the veneer must be provided by tying it to the wood frame. This is accomplished by embedding steel straps in the masonry and nailing the ends of these straps to the frame.

As previously noted the manufacturing of brick has vastly matured since the inception of the masonry concept. Manufacturing processes that include computers to control the temperature of burning of bricks are now used.

The seven major steps in the manufacture of bricks are:

1. Several clays are blended together to obtain a uniform clay mixture.
2. Any non- clay substances are removed.
3. Clay is crushed to powder.
4. Clay powder is combined with water.
5. The wet clay mixture is forced through dies for shaping and cut into appropriate lengths.
6. Bricks are dried.
7. Bricks are burned at temperatures in excess of 2000°F.

The final step of burning the material allows the individual grains to cement together to form a hardened brick.

Clay bricks are available in a variety of sizes to satisfy architectural requirements. Some of the more common sizes that are used are:

Table 5-1—Brick Sizes

Type	**Length (in.)**	**Height (in.)**	**Depth (in.)**
Modular	7-5/8	2-1/4	3-5/8
Norman	11-5/8	2-1/4	3-5/8
Economy	7-5/8	3-5/8	3-5/8
Utility	11-5/8	3-5/8	3-5/8

Using the modular type as a measuring stick, a review of the table shows that a norman brick is 4 in. longer, the economy is the same length but 1.5 in. higher, and the utility has both increases.

The number of bricks needed to construct a 100-ft^2 wall with 3/8-in. mortar joints for the four types presented are:

Modular	686
Norman	457
Economy	450
Utility	300

Bricks are also designated according to the manner in which they are positioned in the wall. A brick that is laid in the customary manner, its length in the long direction of the wall, is called a *stretcher*. When a brick is positioned such that its longest length is into the wall, it is called a *header*. If these above two cases are modified to have the brick laid on its side, the header becomes a rowlock and the stretcher becomes a rowlock stretcher. Figure 5-1 provides illustrations of these various position designations.

Clay bricks can also be designated by material specifications. These can be found in the American Society of Testing Material Standards (ASTM) C62 and C216. C62 provides specifications for bricks where the appearances is of limited importance. C216 is the specification for bricks where appearance is important.

For purposes of this discussion, it is assumed that only bricks not in contact with the earth are used in vertical surfaces where appearance is of interest. Under these circumstances, bricks are placed into one of two categories, SW or MW. SW is the grade designation for bricks subjected to severe weathering. MW is the grade specification for bricks subjected to moderate weathering. Grade SW bricks have higher strength and less of a tendency to absorb water than Grade MW. As a rough rule, Grade SW bricks are required throughout the United States, except for the lower portions of California, Arizona, Texas and Florida.

Thus far, this discussion has labeled bricks by size, position in the wall and exposure type. Clay bricks are also classified by the extent of chippage, dimensional tolerances and distortion. Three classifications are available based on these criteria. These are FBA, FBS and FBX. The most common type of brick used is FBS, which is for general usage. FBA is for uses that require nonuniform sizes

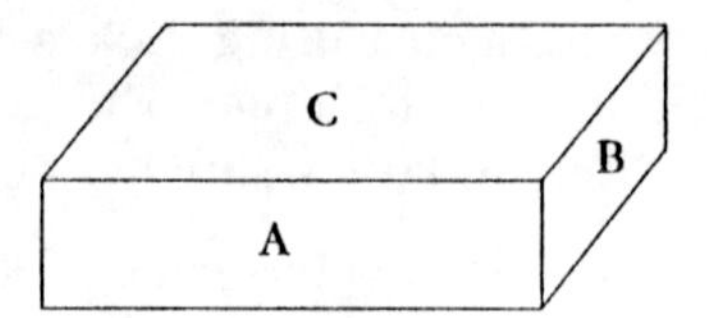

Stretcher

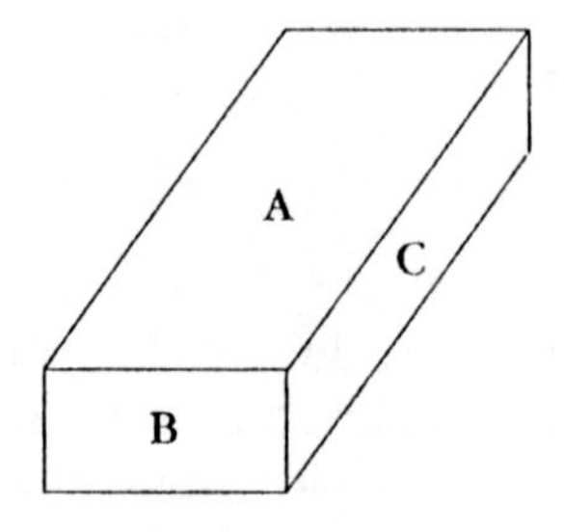

Header

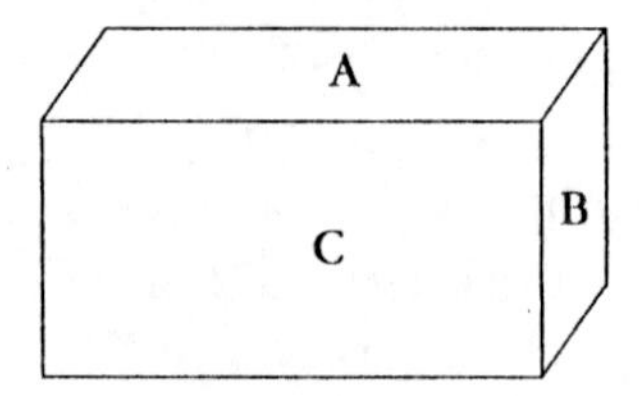

Rowlock Stretcher

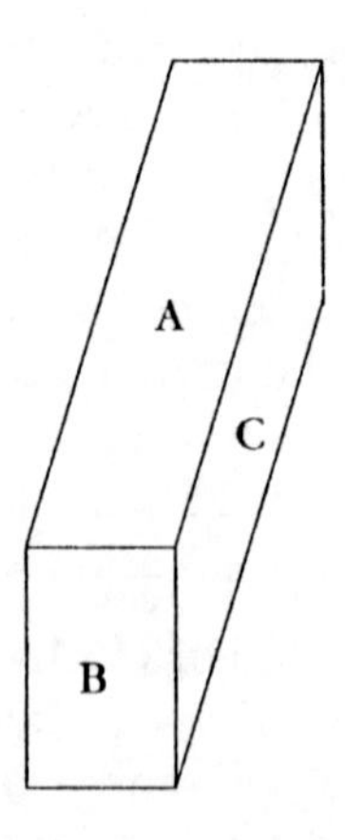

Rowlock

Figure 5-1—Designation of brick by position

or texture. FBX is used in construction where the dimension of the brick must be precise. The requirements for these designations can be found in ASTM C216.

As an example of these designations, consider the following. The dimensions of a brick that is laid on its side is 7-5/8 by 2-1/4 with the latter dimension at the front end. The brick has an irregular shape and texture. If the brick is utilized in St. Louis, Missouri, its classification is:

- Modular–because it has dimensions of 2-1/4 by 7-5/8
- Rowlock–because it is laid on its side
- FBA–because it has irregular texture or shape
- SW–because it is subjected to severe weather conditions

Another widely used masonry product is the concrete block. Clay bricks are burned at high temperature to cause the particles to bind to each other. Concrete blocks use chemical bonding, rather than heating, to bind the individual particles. The same basic mixture as is used in concrete production is utilized to manufacture bricks. Cement, sand and aggregate are mixed to make this product with limits set on aggregate size.

Specifications for load-bearing masonry units are provided in ASTM C90. This specification lists the minimum thickness of the face shell and web thickness for hollow concrete units. For 3-in. wide units, the face thickness shall not be less than 3/4 in. For 6-in. wide units, and larger, the face thickness ranges from 1 to 1-1/2 in.

Some of the nominal sizes in inches for masonry blocks are:

$4 \times 8 \times 16$

$8 \times 8 \times 16$

$10 \times 8 \times 16$

$12 \times 8 \times 16$

The first number is the thickness of the unit, the second is the height and the third is the length.

Walls, just as in concrete slabs and wall construction, can be reinforced or unreinforced. If the blocks are provided with steel reinforcing in the horizontal and vertical direction, the load-carrying capacity of the wall is greatly increased. This increase is the result of the additional tensile capacity provided by the steel since the tensile of the concrete block itself is small.

Although it is only a small percentage of the volume of a masonry wall, the mortar plays just as an important role. If the mortar does not provide the proper bond, the strength of the brick utilized is inconsequential.

Mortars are divided into different types by ASTM C270, Standard Specifications for Unit Masonry. The types that will be discussed are Types M, S, N and O. The types of mortar are distinguished by the ratio of components of cement, lime and aggregate.

Type M

This is the type with the greatest compressive strength (2500 psi at 28 days). It is used in applications where the wall is subjected to significant compressive loads. The proportion of cement/lime/aggregate is 1/0.25/3.5.

Type S

Provides excellent tensile bond strength and is used for applications where there are lateral loads, such as wind, soil, and seismic. The proportions of cement/lime/aggregate is 1/0.5/4.5.

Type N

Used for veneers and load bearing walls. The proportion of cement/lime/aggregate is 1/1/6.

Type O

Used in applications where freezing and thawing is not expected. The proportion of cement/lime/aggregate is 1/2/9.

Advanced Discussion

Masonry walls provide stiffness and strength to resist both vertical and lateral loads. These attributes only exist after the wall has been constructed and the mortar has set. Before the mortar has set, a concrete block wall is merely a pile of blocks rather than a solid wall. During construction, the wall is vulnerable to collapse from any lateral force, particularly wind forces.

Because of its lack of stability during construction, masonry walls should not be constructed when high winds are present or expected. If any significant wind is expected, bracing might be required for temporary lateral support of the wall. Figure 5-2 provides a detail of bracing for masonry walls to resist lateral forces. Because the direction of the wind might change, bracing should be provided on both sides of the wall.

Some publications provide detailed information on when and how to brace walls. Two of these publications are *TEK Bulletin 72* from the National Concrete Masonry Association and documents produced by American Concrete Institute Committee 531.

A rough but conservative approach for determining when temporary bracing is required for unreinforced concrete masonry walls is:

8-in.-thick wall: Use temporary bracing if the wall is taller than 43 ft minus a foot for every mile per hour of peak wind velocity.

10-in.-thick wall: Use temporary bracing if the wall is taller than 50 ft minus a foot for every mile per hour of peak wind velocity.

Example: A wall is being constructed at a location where the peak wind velocity is expected to be 34 mph. What is the maximum height that the wall can be constructed without requiring bracing?

If the wall is 8 in. thick, the height that wall bracing will be required can be conservatively taken as:

$$\begin{aligned} \text{Height} &= 43 - \text{peak wind speed} \\ &= 43 - 34 \\ &= 9 \text{ ft} \end{aligned}$$

If the wall is 10 in. thick, the height that bracing will be required can be conservatively taken as:

$$\begin{aligned} \text{Height} &= 50 - \text{peak wind speed} \\ &= 50 - 34 \\ &= 16 \text{ ft} \end{aligned}$$

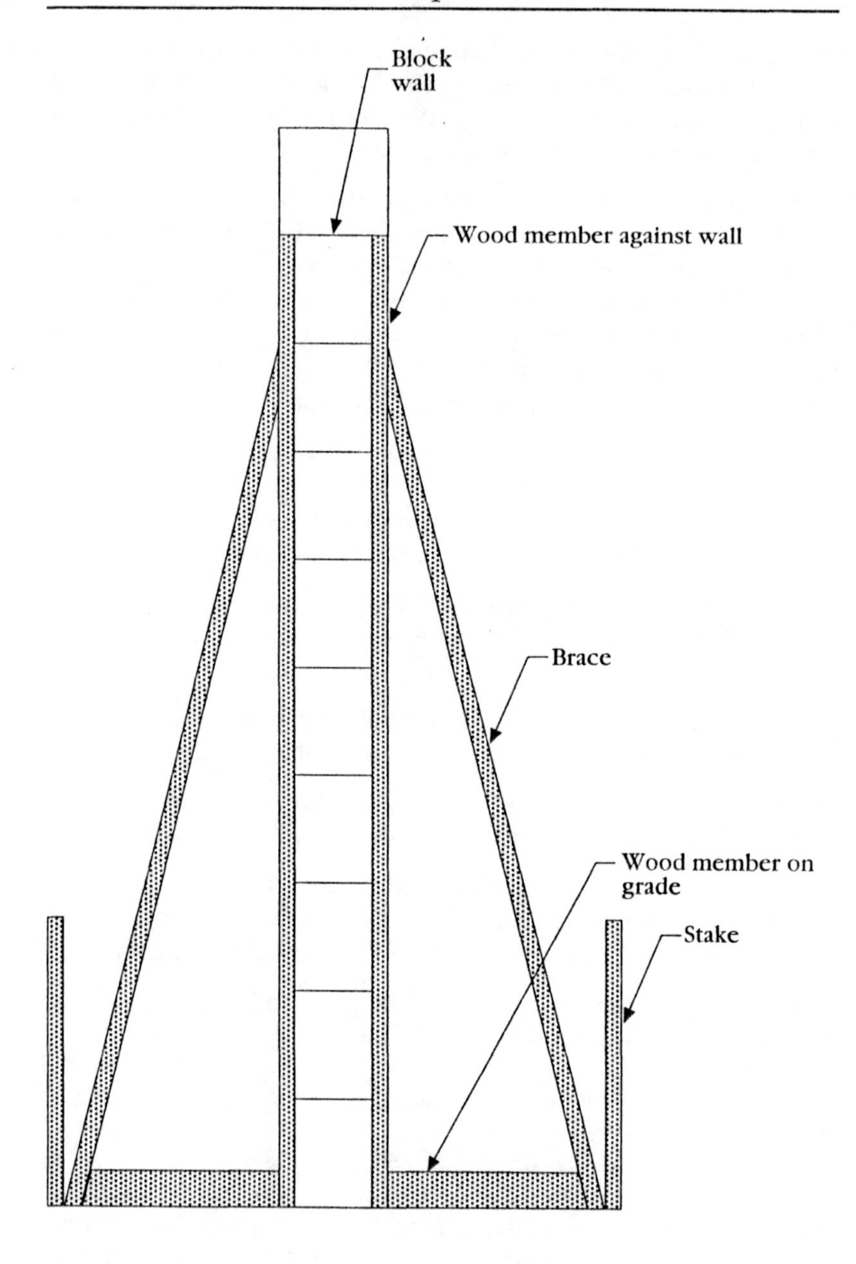

Figure 5-2—Temporary bracing of block wall

The rule-of-thumb equations are conservative for low-speed peak winds. However, for low-speed winds, the height of the wall that can be constructed without requiring bracing is in excess of what is used for residential construction.

The rule-of-thumb equations are also conservative for high-speed peak winds. In fact, bracing would be required when wind speeds are greater than 43 mph for an 8 in. block wall and 50 mph for a 10 in. block wall. Calculations show that this is not necessary, but nevertheless, when the speeds of this magnitude are expected, a rule-of-thumb method might not be appropriate.

Loads & Codes

A success construction project can only be accomplished if the design of the house has been done properly. Similarly, a proper design cannot be accomplished unless the loads that the structure will be subjected to are known. In fact, determining what loads will be applied to the structure is one of the first tasks facing the designer.

The loads applied to a structure are a function of where the structure is located geographically, what the function of the structure is, and the type of material used to build the structure.

The wind speeds that a building must be designed for might not exceed 70 mph in Kentucky but are in excess of 100 mph in some cities in Florida. Similarly, snow loads in Maine might produce loads of 100 lb/ft^2 but are nonexistent in Florida.

The first chapter in the Loads and Codes portion of this text presents codes, specifications and recommended practices that are common in the residential construction industry, as well as other topics of interest. The following three chapters discuss the actual determination of the loads for the structure's own weight, loads applied by the occupants and by snow and wind effects.

6 Codes, Specifications and References

Introduction

The requirements for a construction project are contained in the contract documents. The contract documents for typical construction projects include:

- Drawings
- Project specifications
- Supplemental specifications
- Additional specifications
- Agreement
- Bidding requirements
- Bonding forms

These contract documents can be quite voluminous. The contract documents for residential construction are not as detailed and will usually include at most, the following:

- Drawings
- Project specifications

Whether the construction project is a multi-million dollar venture or a small residential project, the contract documents will never be complete. For example, the contract documents can list the requirements of what concrete testing will be necessary. But it is not practical for the project documents to provide complete information on how the test should be performed, how many samples should be taken, how to calculate standard deviations of the results, etc. It is not practical for the project documents to provide every detail on the requirements for electricity, plumbing, structural loading, safety and a multitude of other issues.

Project documents will refer to codes, government documents, specifications of documents prepared by government agencies, organizations existing for the purpose of writing codes, and trade association recommended practices. Reference in the contract documents to these bodies of work require the contractor to provide a structure that meets more requirements than detailed in the contract documents.

This chapter will provide a list of some of the codes, recommended practices and specifications used in the residential construction industry. The entity that makes the publication available is listed. The topics in this chapter are:

- Safety
- Building codes
- Wood
- Concrete
- Steel
- Masonry
- Materials

Safety

Occupational Safety and Health Administration Standards
(OSHA)
United States Department of Labor
Washington, D.C.

Covers safety requirements for the construction industry. Some of the topics provided in OSHA that are applicable to the residential construction industry include: excavation, ladders, fall protection, openings in floors and scaffolds, etc.

Accident Prevention Manual
The Associated General Contractors of America
Washington, D.C.

Provides safety requirements in the construction industry. Covers topics very similar to those provided in the Occupational Safety and Health Administration Standards (OSHA). This text is written in a user-friendly language and is a helpful quick reference guide.

Handling and Erecting Wood Trusses
Truss Plate Institute
Madison, Wisconsin

A short publication that provides very important information for contractors that utilize prefabricated and job-built trusses. Discusses the process of proper erection procedures for trusses to avoid construction collapses and injuries. Information is also given on temporary bracing and lifting techniques.

Worker Safety on Pitched Roof
American Plywood Association
Tacoma, Washington

This one-page publication presents the factors that contribute to accidents. More importantly, it provides safety recommendations on how to avoid these accidents.

Codes

Basic Building Code (BOCA)
Building Officials Code Administration International, Inc.
Country Club Hills, Illinois

One of the three major building codes used in the United States. Predominately used in the Midwest. Provides the minimum code requirements for design, construction, quality of materials, use and occupancy and other related requirements for buildings.

Uniform Building Code (UBC)
International Conference of Building Officials
Whittier, California

Another of the three major building codes used in the United

States. Predominately used in the West. Provides the minimum code requirements for design, construction, quality of materials, use and occupancy and other related requirements for buildings.

Standard Building Code (SBC)
Southern Building Code Congress International
Birmingham, Alabama

The third of the three major building codes used in the United States. Predominately used in the South. Provides the minimum code requirements for design, construction, quality of materials, use and occupancy and other related requirements for buildings.

CABO: One and Two Family Dwelling Code Joint effort by the
Building Officials Code Administration International, Inc.,
International Conference of Building Officials and the
Southern Building Code Congress International

A joint effort of the three major building code-writing organizations. Provides the minimum requirements for design, construction, quality of material, use and occupancy for detached one-or-two family dwellings, not more than three stories in height. Utilizes code requirements from all three codes.

Minimum Design Loads for Buildings and Other Structures
(ASCE Standard 7-95)
American Society of Civil Engineers
New York, New York

Provides the provisions for determining the loads required to design a structure, including snow, wind and seismic loads. Maps are provided that depict the wind, seismic and snow intensities throughout the United States.

National Electric Code
Sponsored by the National Fire Protection Association
Quincy, Massachusetts

Details requirements for wiring methods, wiring materials and equipment. Also provides rules for wiring of special equipment, including swimming pools, stairway chair lifts and sound-recording equipment.

Plumbing Codes
Provided by each of the Code Writing Organizations

Basic Plumbing Code by the Building Officials Code Administration International, Inc. (Illinois).

ICBO Plumbing Code by the International Conference of Building Officials (California).

Standard Plumbing Code–Southern Building Code Congress International (Alabama).

Wood

National Design Specifications for Wood Construction
National Forest Products Association
Washington, D.C.

Specifications that present the design requirements for wood beams, columns and other structural members. Provides design considerations for glued laminated timber, wood fasteners and panel products.

Supplement to Design Specifications for Wood Construction
National Forest Products Association
Washington, D.C.

This supplement provides design values for structural lumber and glued laminated timber. This supplement is a necessity to fully utilize the National Design Specifications.

APA Design/Construction Guide
American Plywood Association
Tacoma, Washington

Reference manual that contains in depth information on plywood. Thorough explanations of grade specifications, allowable span lengths, and building code requirements for plywood. Explicitly detailed drawings show fire-resistant construction for wood and panel assemblies.

APA Residential Guide
American Plywood Association
Tacoma, Washington

A 4 × 7 in. spiral-bound pocket reference manual that provides summarized information of the APA Design/Construction guide. The usefulness of this information is such that it is advisable to keep it in a tool box or brief case of the homebuilder.

House Building Basics
American Plywood Association
Tacoma, Washington

A practical primer for new contractors on topics such as: Lay out of the foundation, floor framing, subfloors and roof framing. Useful information on the basics of nail types and sizes.

Installation and Preparation of Plywood
Underlayment for Resilient Floor Covering
American Plywood Association
Tacoma, Washington

This six-page pamphlet provides recommendations for underlayment for tile and sheet flooring. Recommends plywood grades for underlayment and fastening schedules.

Steps to Constructing a Solid, Squeak-Free Floor System
American Plywood Association
Tacoma, Washington

This two-page pamphlet outlines the four steps to a squeak-free floor.

Concrete

Cement and Concrete Terminology (ACI 116)
American Concrete Institute
Detroit, Michigan

Contains definitions for more than 2000 terms pertaining to concrete construction. Word definitions are easy to understand.

Building Code for Concrete Structures (ACI 318 & 318.1)
American Concrete Institute
Detroit, Michigan

Extremely thorough code and commentary for the design of concrete buildings. One version is for reinforced concrete (ACI

318) and the other is for unreinforced concrete (ACI 318.1). These publications provide detailed information from material requirements to seismic design.

Notes on ACI 318–Building Code for Concrete Structures
Portland Cement Association
Skokie, Illinois

Provides in-depth explanation of the requirements of ACI 318. Detailed examples illustrate the application of a vast majority of the code.

The Homeowners Guide to Building with Concrete,
Brick and Stone
Rodale Press
Emmaus, Pennsylvania

An excellent text that discusses the fundamentals of concrete for residential construction. A multitude of illustrations that assist in explaining important concepts. Details provided can be actually implemented by the builder.

Concrete Homebuilding Systems
McGraw-Hill
New York, New York

State-of-the-art textbook providing a complete overview of residential structures built with concrete. Various concrete building systems are detailed, including a directory of product manufacturers. Useful information regarding costs, market appeal and logistics are also provided.

Guide to Residential Cast-in-Place Concrete Construction
(ACI 332)
American Concrete Institute
Detroit, Michigan

Provides complete overview of concrete topics pertaining to the residential construction industry. Topics covered include formwork, proportioning of concrete and repair of surface defects. Reinforcing bar layout and joint details are provided.

Steel

Manual of Steel Construction
American Institute of Steel Construction
Chicago, Illinois

This book provides geometric properties for steel angles, channels, wide flange sections and other standard members. Information is presented on beam and column design. Welded and bolted connection information is included.

Specifications for the Design, Fabrication
and Erection of Steel Buildings
American Institute of Steel Construction
Chicago, Illinois

Guidelines for all phases of steel building construction. Useful text provides easy to understand information that supplements requirements of the *Manual of Steel Construction*.

Masonry

Building Code Requirements for Masonry Structures
(ASCE 5-88)
American Society of Civil Engineers
New York, New York

Presents the requirements for design and construction of masonry structures. Topics include requirements for strength, serviceability and quality assurance. Includes rules for wall and pilaster design.

Specifications for Masonry Structures (ASCE 6-88)
American Society of Civil Engineers
New York, New York

Provides information on inspection, testing and placement of reinforcing, strength determination and provisions for grouting.

Concrete Masonry Handbook
Portland Cement Association
Skokie, Illinois

This easy-to-read text is for builders, architects and engineers. This book clearly and thoroughly covers the fundamentals of masonry construction, including discussions on materials.

Masonry Design and Detailing
McGraw-Hill
New York, New York

Comprehensive, as well as interesting, textbook on all areas of masonry. Glossary is provided at end of text.

Material

Standard Definitions of Terms Relating to Wood (ASTM D 9)
American Society for Testing and Materials
Philadelphia, Pennsylvania

A collection of the more general terms that are common to the wood industry. Provides references where more specialized definitions may be located.

Standard Definitions of Terms Relating to Veneer and Plywood (ASTM D 1038)
American Society for Testing and Materials
Philadelphia, Pennsylvania

A collection of the general terms that are common to plywood applications.

Standard Terminology Relating to Concrete and Concrete Aggregates (ASTM C 125)
American Society for Testing and Materials
Philadelphia, Pennsylvania

A collection of the general terms that are common to concrete construction.

Standard Specifications for Structural Steel (ASTM A 36)
American Society for Testing and Materials
Philadelphia, Pennsylvania

Specifications for various steel shapes for use in buildings, bridges and general purposes.

Standard Specifications for Concrete Aggregates (ASTM C 33)
American Society for Testing and Materials
Philadelphia, Pennsylvania

Defines the requirements for grading and quality of fine and course aggregate for use in concrete. The specifications can be used by designers to define the quality to be used or by contractors as a purchase document.

Standard Specifications for Ready-Mixed Concrete (ASTM C 94)
American Society for Testing and Materials
Philadelphia, Pennsylvania

Covers requirements for concrete delivered to purchaser in a freshly mixed and unhardened state. Does not cover placement, finishing or curing of the concrete.

Advanced Discussion

This section provides information similar to that provided in the preceding discussion on this topic. However, this section provides textbooks and similar information that provide technical details and applications of the aforementioned topics. The majority of these textbooks are geared more toward engineers and architects than general contractors. Nevertheless, there is no reason that an ambitious contractor could not obtain useful information from these documents.

Safety

Excavation Safety
The Aberdeen Group
Addison, Illinois

A well-written and detailed text book providing an explanation of the excavation portion of the Safety and Health Administration Standards (OSHA).

Codes

Guide to the Use of the Wind Load Provisions
of ASCE Standard 7-88
American Society of Civil Engineers
New York, New York

In-depth discussion on the wind provisions in ASCE Standard 7-88. Examples are provided to illustrate the necessary calculations to determine the magnitude and distribution of wind forces.

Wood

Design of Wood Structures
McGraw-Hill
New York, New York

One of the most popular technical design books for wood construction. Comprehensive discussion of design of beams and columns under a variety of loads. Includes information on the design of glued laminated beams.

Concrete

Design and Control of Concrete Mixtures
Portland Cement Association
Skokie, Illinois

One of the most informative and well-written texts on the behavior and properties of concrete. Written so that only limited technical background is necessary to understand topics such as: Proportioning mixtures, placing, finishing and curing concrete. Each chapter has extensive references for those looking for more information.

Cold Weather Concreting (ACI 306)
American Concrete Institute
Detroit, Michigan

Information is provided to assist in producing a strong structure when it is constructed under cold-weather conditions. Discusses preparation and protection during curing.

Hot Weathering Concreting (ACI 305)
American Concrete Institute
Detroit, Michigan

Less-than-acceptable results might occur when concrete is placed at high temperatures. Proper precautions are necessary to prevent cracking, particularly during the curing stages. Information is provided that discusses the effects of high temperatures and precautions to avoid problems.

Guide for Concrete Floors and Slab Construction (ACI 302)
American Concrete Institute
Detroit, Michigan

This publication is more directed toward floors that have loads in excess of the norm for residential construction. However, the information provided on proper preparation and installation is very pertinent to residential construction. Information is also provided on common causes of floor and slab failures.

Prediction of Creep, Shrinkage and Temperature Effects
(ACI 209)
American Concrete Institute
Detroit, Michigan

Provides analytical methods for determining the effects of sustained loads, moisture changes and temperature on concrete structures. Although creep and shortening of members from long-term loading is extremely unusual in residential construction, the other issues might not be so rare. The beginning of this publication provides a good overview of these potential changes in volume.

Steel

Structural Welding Code
American Welding Society
Miami, Florida

Welding requirements for all types of structures when conventional welding techniques are utilized. Commentary provides explanation and insight into code requirements.

Steel Structures: Design and Behavior
Harper Collins
New York, New York

Excellent engineering textbook that provides full details of steel-structure design. This textbook is not written for residential construction. However, it is a great resource for any person considering designing homes constructed with steel framing.

Masonry

Reinforced Masonry Engineering Handbook
Masonry Institute of America
Los Angeles, California

This text provides the fundamentals of the structural design of masonry members. Design aids are included. An engineering background is needed to understand the majority of this text.

Materials

Standard Practice for Establishing Structural Grades and Related Allowable Properties for Visually Graded Lumber (ASTM D 245)
American Society for Testing and Materials
Philadelphia, Pennsylvania

Covers the basic principles for nonmechanical grading of lumber and establishing the strength and stiffness. Basically determines the structural qualities of a piece of lumber after taking into account reductions for knots and cross-grain loading.

Standard Test Methods For Mechanical Properties of Lumber and Wood-Base Structural Materials (ASTM D 4761)
American Society for Testing and Materials
Philadelphia, Pennsylvania

Test methods to determine the strength and related properties of wood members.

Standard Test Methods for Establishing Clear Wood Strengths (ASTM D 2555)
American Society for Testing and Materials
Philadelphia, Pennsylvania

Provides information on providing strength characteristics of the optimal piece of lumber. Presents the factors for consideration in the adjustment of the optimal values for nonoptimal lumber.

Standard Test Methods for Mechanical Fasteners in Wood
(ASTM D 1761)
American Society for Testing and Materials
Philadelphia, Pennsylvania

Presents methods of conducting tests for nails, staples, screws, bolts and timber connectors.

Standard Test Method for Establishing Stresses for Structural
Glued Laminated Timber (ASTM D 3737)
American Society for Testing and Materials
Philadelphia, Pennsylvania

Test methods to determine properties for bending, tension, compression, shear and modulus of elasticity for glued laminated timber.

Standard Test Method for Resistance of Concrete to Rapid
Freezing and Thawing (ASTM D 666)
American Society for Testing and Materials
Philadelphia, Pennsylvania

Tests to determine the effects of variations in properties and conditioning of concrete in the resistance to freezing and thawing for a particular application.

7

Dead and Live Loads

Introduction

Structures must be designed to withstand the forces produced by a wide range of loading conditions, including:

- Dead loads
- Live loads
- Snow loads
- Wind loads
- Soil loads
- Earthquake loads
- Thermal loads

The first five in this list are the most common loads for residential construction. Dead and live loads for the floors are discussed in this section. Dead loads are the weight of permanent objects (walls, floors, plaster, carpet, etc.) Live loads are the forces produced by transient objects, such as the movement of people, objects, vehicles, bikes, etc.

It is interesting to note that live loads may be the predominant force in some cases or a small force in others. In bridge design the dead load of the structure is a small percentage of the overall load since a line of trucks or cars (i.e., the live load) packing a bridge is a significant load. Buildings on the other hand are predominately dead load. The weight associated with the structure of a high-rise building can significantly exceed the live-load forces produced by the occupants.

The building codes define what is the minimum live load that a structure must be designed to withstand. The loads consist of a uniformly distributed load and possibly a concentrated load. A sample of the typical minimum uniformly distributed load that must be designed for is:

Balcony for one- or-two family dwelling	60 psf
Hospitals	40 psf–80 psf
Libraries–reading room	60 psf
Libraries–stack room	150 psf
Light manufacturing	100 psf
Heavy manufacturing	250 psf
Hotels	40 psf–100 psf
School–classroom	40 psf
School–corridors	80 psf
Residential–attic	20 psf
Residential–nonsleeping rooms	40 psf
Residential–sleeping rooms	30 psf
Stores	75 psf

In addition to the uniformly distributed live loads discussed previously, floors of buildings must also support a concentrated load under some circumstances. These concentrated loads shall be placed so as to produce maximum stresses in the member being designed. Some of the concentrated loads are:

Hospitals	1000 lb
Library	1000 lb
Schools	1000 lb
Manufacturing	2000 lb

The minimum design live loads are prescribed by the code. Dead loads are not prescribed by code but rather use the actual weight of the member. The dead load of some material for residential construction as well as general interest are:

Hardwood flooring per in. thickness	4 psf
Underflooring per in. thickness	3 psf
Linoleum	2 psf
Wood joists—2 × 12 at 16 in.	3 psf
Wood joists—2 × 12 at 12 in.	4 psf
Wood joists—2 × 8 at 16 in.	2 psf
Wood joists—2 × 8 at 12 in.	3 psf
Concrete	150 psf
Dry dirt	100 psf
Wet dirt	120 psf
Wood	35-50 psf
Asphalt shingles	2 psf
Wood shingles	3 psf
Stud walls	4 psf
Stud walls-drywall	10 psf

Advanced Discussion

It is highly unlikely that a large floor area would ever be subjected to the live load forces it is designed for. Building codes recognize that it would be overkill if a structural member were designed to support the live loads over a large area. For this reason codes allow for a live-load reduction. A reduction in the live load is allowed for any member that supports a tributary load area in excess of 200 square feet.

An example of some of the reduced live-load percentages are:

For beams:

$$\text{Live-load percentage} = 0.25 + 10.6/\sqrt{A}$$

For Columns:

$$\text{Live-load percentage} = 0.25 + 7.5/\sqrt{A}$$

where A is the tributary area.

Example 1: A steel beam supports the first floor joists of a residential structure. The steel beam is 12-ft long and is 15-ft away from the adjacent concrete walls.

The tributary area is 12 ft × 15 ft for a square footage of 180 ft^2. Therefore, no live load reduction is allowed.

Example 2: Same as Example 1, except that the beam is 14 ft long. The tributary area is 14 × 15 for a total of 210 ft^2. The reduction factor is:

$$\begin{aligned}\text{Reduction factor} &= 0.25 + 10.6/\sqrt{210}\\ &= 0.98\end{aligned}$$

Therefore, if the design live load is 50 psf, it could be reduced to 49 psf (0.98 3 50 = 49 psf).

Live-load reductions are not allowed to exceed 50% for members supporting one floor nor 40% for members supporting more than one floor. These live-load reduction factors are only applicable to floor loads since roof live loads have a different live-load reduction factor.

As it is unlikely that a floor will be subjected to the live load it is designed for, it is also unlikely that it would ever be fully subjected to all possible loads at one time. For example, it would be unlikely that a column would be subjected to full wind load, snow load and live load at the same instant in time. The building codes handle this unlikely situation by providing load-reduction factors when loads are being designed for simultaneously as follows:

- If design is for dead load, live load, wind load and snow load, ignore one-half of snow load.
- If design is for dead load, live load, wind load and snow load, ignore one-half of wind load.

8

Roof Loads

Introduction

The roof of a house must be designed to support its own weight, wind loads, live loads and snow loads. The structure's own weight as well as wind forces are discussed in other chapters. This topic, roof loads, covers roof live loads and snow loads.

Roof live loads can come from workers placing shingles on the roof, water on the roof or the homeowner making an inspection. Live loads can also result from the temporary placement of construction materials on the roof. The minimum load that a roof must be designed for is a function of the slope of the roof. The following are the minimum live loads based on the 1993 BOCA National Building Code.

Roof angle less than 19 degrees	20 psf
Roof angle between 19 and 45 degrees	16 psf
Roof angle in excess of 45 degrees	12 psf

It should be noted that the roof live load is based on the square footage of the horizontal dimensions of the roof. This is explained in Figure 8-1. Each of the roofs shown cover the same size house. But, each will have a different distributed load since the roof angles are different, although the loads will be distributed over the exact same area for all three cases. Compare the results of Figure 8-1 with those of Figure 8-2. In Figure 8-2, the overall load is equal for

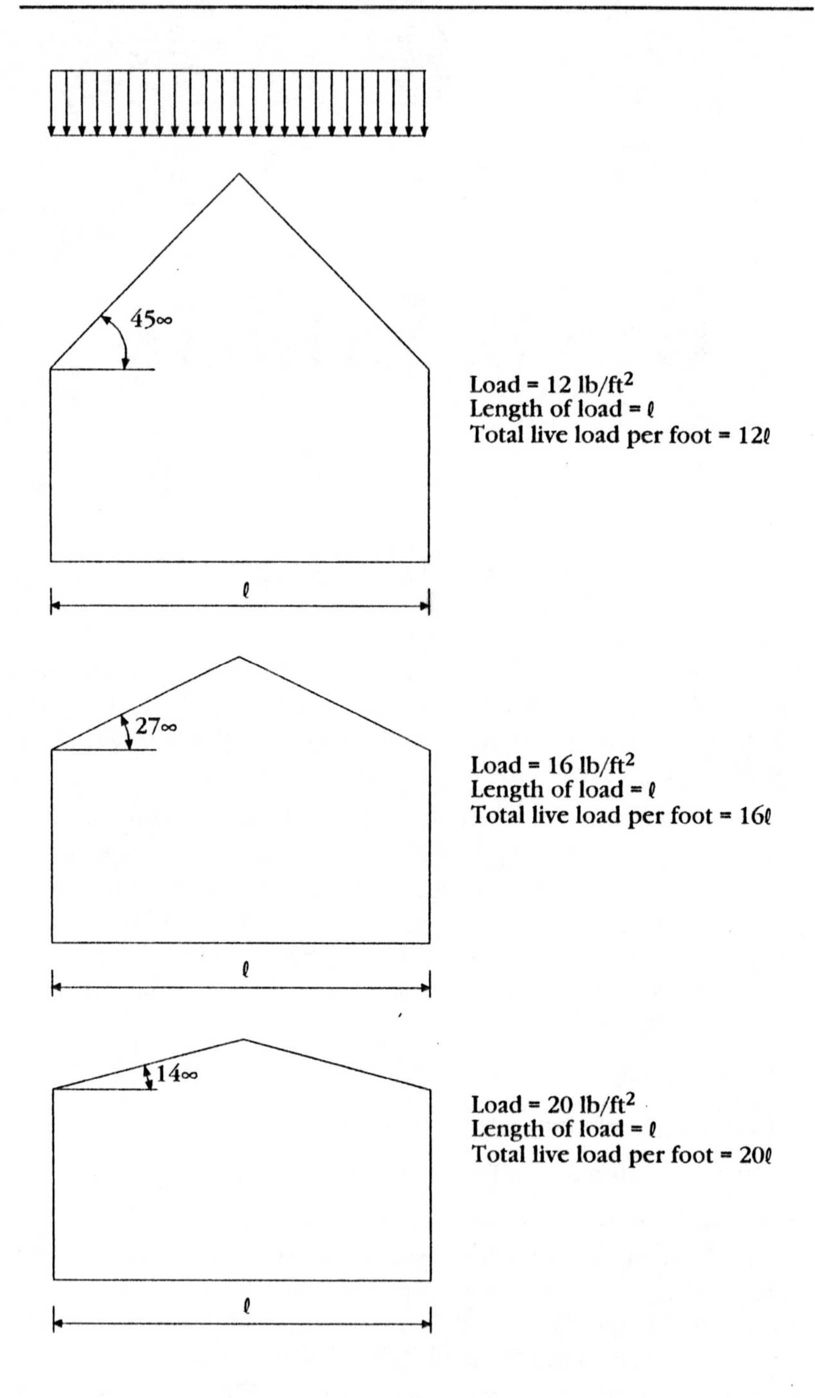

Figure 8-1 Tributary area of live load equal for all cases

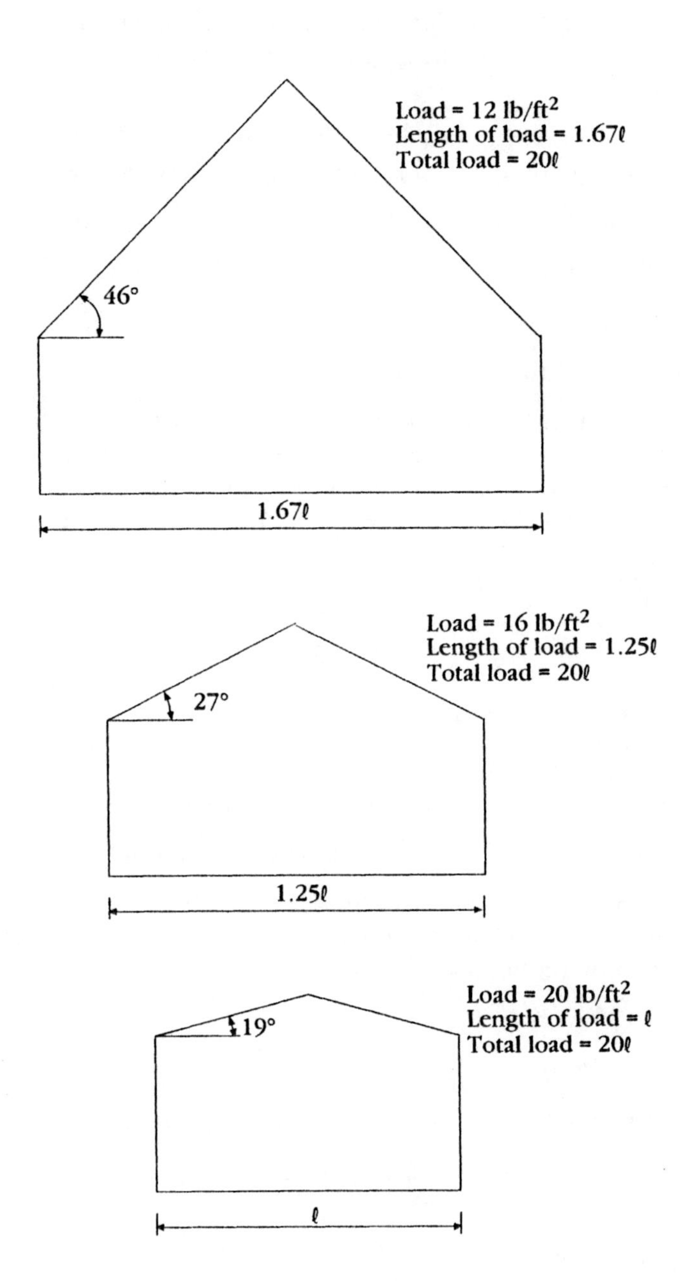

Figure 8-2 Total load on roof equal for all cases

all cases. However, the area it is spread over differs. These cases illustrate that the steeper the roof, the less live load need be designed for.

The other load to be discusses in this section is snow load. The amount of snow load on a roof is a function of the following:

- Geographic location
- Slope of the roof
- Local terrain
- Importance of the structure

The amount of snow that would be expected obviously depends on the local climate. The expected ground snow load in Florida is zero. In some parts of Maine it can be as much as 100 lb/ft^2. The ground snow loads that can be expected in the various areas of the United States are as follows:

Northeast	15 psf–100 psf
Southeast	0 psf–15 psf
Northern Midwest	20 psf–70 psf
Southern Midwest	0 psf–20 psf
Northwest	Extreme Local Variation
Southwest	Extreme Local Variation

As was the case for live loads, the snow load is also a function of the roof slope. For a roof with a slope in excess of 30 degrees, the snow load can be reduced as follows:

35 degrees	12.5%
40 degrees	25.0%
45 degrees	37.5%
50 degrees	50.0%
55 degrees	62.5%
60 degrees	75.0%
65 degrees	87.5%
70 degrees	100%

The building codes require that certain structures have increased or decreased snow loads based on the use of the structure. The changes in snow loads are as follows:

Essential facilities	+20%
Gathering places (more than 300 people)	+10%
Low-hazard facilities	-20%

The importance factors do not affect residential construction.

The final factor having a bearing on the snow loads is the local terrain. The terrain factor is 0.7 for all structures with two exceptions. If a structure is located in a densely forested area or similar to that, the terrain factor is 0.9. If the structure is in open terrain, the factor is 0.6.

Example: A residential structure located in Central Illinois has a roof at a 35-degree angle. The structure is located in a sheltered area. Determine the appropriate snow and roof live load.

The ground snow load in Central Illinois is 20 lb/ft^2. A 35-degree angle roof has a reduction factor of 12.5%. A structure located in a sheltered terrain has a terrain factor of 0.9.

Roof snow load = Ground snow load × roof slope factor × terrain reduction factor

$$= 20 \times (1 - 0.125) \times 0.9$$
$$= 15.75 \text{ lb/ft}^2$$

The roof live load for a roof with a slope between 19 and 45 degrees is 16 psf. Therefore, roof live load is larger.

Advanced Discussion

Structural failures from snow loads are not always the result of the snow uniformly accumulating on the roof until the roof collapses. The failures are often the results of drifting snow causing increases in the load, increased loads on the lower roofs caused by snow sliding from higher adjacent roofs and by unbalanced snow loads increasing stresses.

Drifting loads occur in the situation shown in Figure 8-3. Drifts might also occur in structures that are not connected but are still close enough for extra accumulation on the lower roof as it blows off the nearby taller structure. Drifting need not be considered when the structures are more than 20-ft apart.

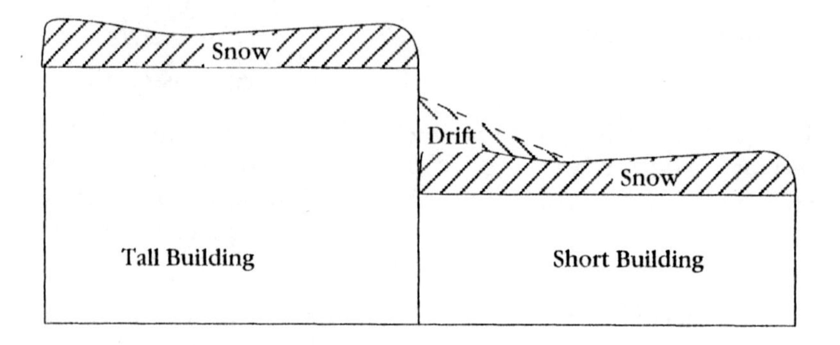

Figure 8-3—Snow drift on roof

Another potentially severe loading condition caused by different height structures in close proximity is snow sliding. Snow sliding occurs in the situation shown in Figure 8-4. Snow sliding forces must be considered if the higher roof has a slope of 20 degrees or larger.

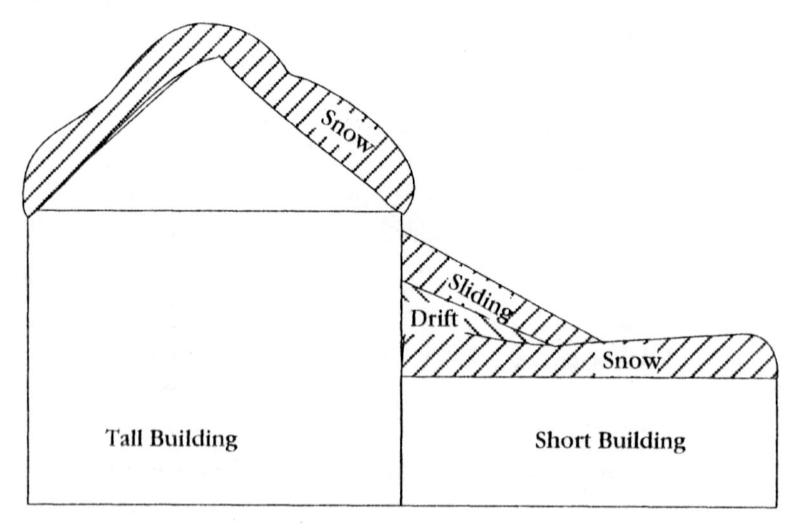

Figure 8-4—Sliding snow and snow drift on roof

The loading from the drifting snow is calculated by determining the weight of the triangle shown in Figure 8-5. Tables 8-1 through 8-6 provide the length, height and density of the snow load for drifting and sliding conditions for ground snow loads of 15 psf to 90 psf.

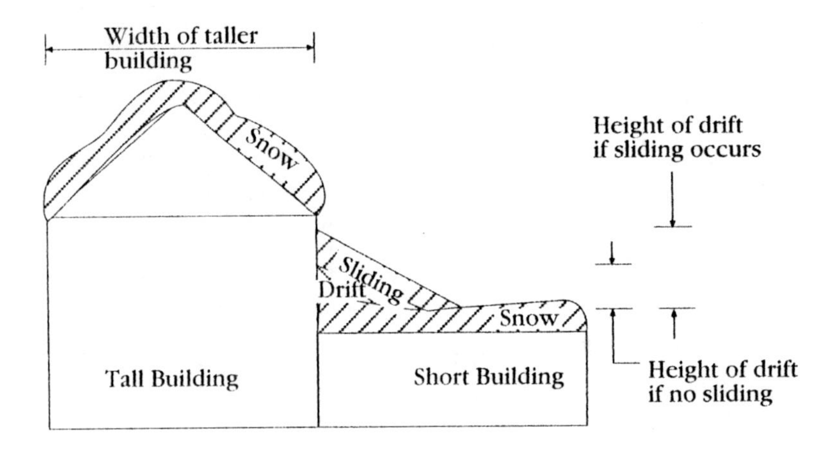

Figure 8-5—Dimensions of snow added from sliding and drifting

The wind can cause the snow load to be removed or reduced on one side of the roof. This unbalanced snow load can create undesirable stresses. Unbalanced snow should be considered for any roof, which is not flat, and has an angle less than 71 degrees. For a roof with a slope up to 20 degrees, half of the load on one side of the roof should be removed and the corresponding stresses reviewed. For slopes greater than 20 degrees but less than 71 degrees, the snow on one side of the roof should not be included in the analysis and the other side should be increased by 25%.

Table 8-1—Snow Drifting and Sliding Ground Snow Load of 15 psf

Width of Taller Building (ft)	Height of Drift if No Sliding (ft)	Height of Drift with Sliding (ft)	Length of Drift (ft)	Density of Snow (psf)
10	0.6	0.8	2.3	16.0
20	1.1	1.55	4.4	16.0
25	1.3	1.8	5.2	16.0
30	1.5	2.1	5.95	16.0
40	1.8	2.5	7.15	16.0
50	2.0	2.86	8.2	16.0
60	2.3	3.2	9.1	16.0

Table 8-2—Snow Drifting and Sliding Ground Snow Load of 30 psf

Width of Taller Building (ft)	Height of Drift if No Sliding (ft)	Height of Drift with Sliding (ft)	Length of Drift (ft)	Density of Snow (psf)
10	0.8	1.16	3.3	17.9
20	1.4	2.0	5.7	17.9
25	1.7	2.3	6.6	17.9
30	1.9	2.6	7.4	17.9
40	2.2	3.1	8.8	17.9
50	2.5	3.5	9.9	17.9
60	2.7	3.8	10.9	17.9

Table 8-3—Snow Drifting and Sliding Ground Snow Load of 45 psf

Width of Taller Building (ft)	Height of Drift if No Sliding (ft)	Height of Drift with Sliding (ft)	Length of Drift (ft)	Density of Snow (psf)
10	1.0	1.4	4.1	19.9
20	1.7	2.4	6.7	19.9
25	1.9	2.7	7.7	19.9
30	2.1	2.99	8.55	19.9
40	2.5	3.5	10.0	19.9
50	2.8	3.9	11.26	19.9
60	3.1	4.3	12.3	19.9

Table 8-4—Snow Drifting and Sliding Ground Snow Load of 60 psf

Width of Taller Building (ft)	Height of Drift if No Sliding (ft)	Height of Drift with Sliding (ft)	Length of Drift (ft)	Density of Snow (psf)
10	1.2	1.7	4.7	21.8
20	1.9	2.63	7.5	21.8
25	2.1	3.0	8.5	21.8
30	2.4	3.31	9.46	21.8
40	2.75	3.86	11.0	21.8
50	3.1	4.3	12.3	21.8
60	3.37	4.72	13.5	21.8

Table 8-5—Snow Drifting and Sliding Ground Snow Load of 75 psf

Width of Taller Building (ft)	Height of Drift if No Sliding (ft)	Height of Drift with Sliding (ft)	Length of Drift (ft)	Density of Snow (psf)
10	1.3	1.8	5.25	23.8
20	2.0	2.86	8.2	23.8
25	2.3	3.2	9.3	23.8
30	2.6	3.6	10.2	23.8
40	3.0	4.2	11.86	23.8
50	3.3	4.6	13.2	23.8
60	3.6	5.06	14.4	23.8

Table 8-6—Snow Drifting and Sliding Ground Snow Load of 90 psf

Width of Taller Building (ft)	Height of Drift if No Sliding (ft)	Height of Drift with Sliding (ft)	Length of Drift (ft)	Density of Snow (psf)
10	1.4	2.0	5.7	25.7
20	2.2	3.1	8.76	25.7
25	2.5	3.5	9.9	25.7
30	2.7	3.8	10.9	25.7
40	3.15	4.4	12.6	25.7
50	3.5	4.91	14.0	25.7
60	3.8	5.35	15.29	25.7

9

Wind Loads

Introduction

Wind forces are not only extremely powerful, they are highly unpredictable. Snow loads and loads produced by the structure's own weight can possibly be fairly accurately predicted by making some relatively simple calculations.

Wind loads cannot be predicted easily. These loads depend on such factors as the local terrain and the building shape. In fact, with respect to high-rise structures, some building officials will not accept typical calculations but require a wind-tunnel test instead. In these special cases, the designer must construct a scaled-down model of the buildings in the surrounding area and also of the proposed structure and place the model in a wind tunnel to accurately determine the forces.

Wind forces are much less complex for one-story and two-story residential structures. However, partial or total failures from wind forces are not uncommon. A large portion of these failures can be avoided if the structure is properly anchored at critical locations.

Figure 9-1 shows the distribution of wind forces on a residential structure. It is important to note two items in this figure. First, a wind force on one side of the structure produces a suction force on the other. This suction force will substantially increase the overall force on the building. The second point to note is that the roof is subjected to an upward pull from the cross wind.

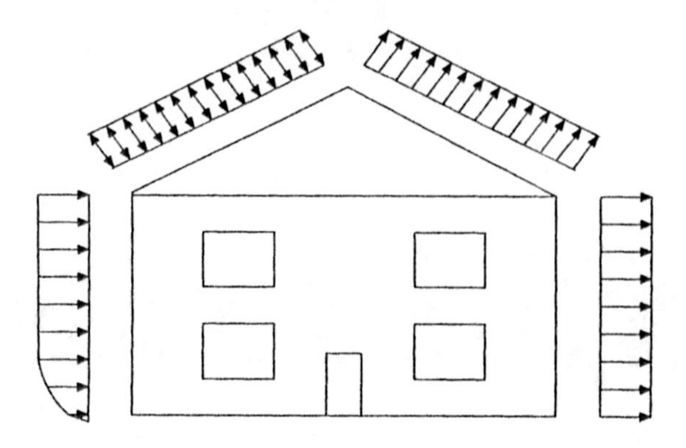

Figure 9-1–Resultant forces on residential structure from wind forces

Three types of failures that can be caused by wind forces are shown in Figures 9-2 through 9-4. Figure 9-2 depicts a failure that is the result of an inadequate foundation connection. A structure should be tied to the foundation by steel bolts that are embedded in the foundation wall. If there is an inadequate amount of bolts, the bolts are improperly sized or are not properly embedded in the concrete wall, the house will lift or slide off the foundation.

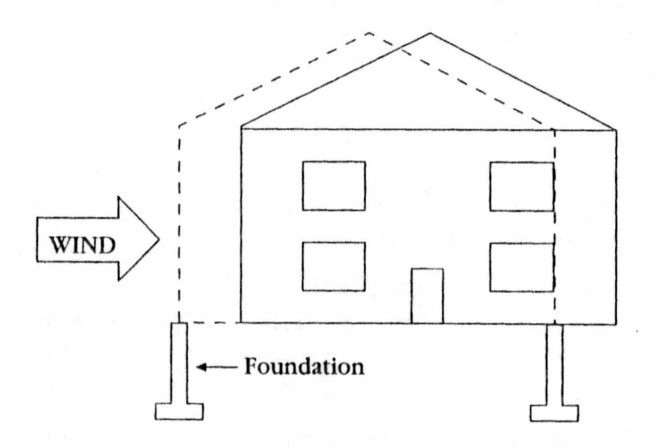

Figure 9-2–Failure of connection of frame to foundation

Figure 9-3 depicts a failure that is similar in nature to the failure discussed in Figure 9-2. In this case, the roof of the structure is not properly tied to the walls. This problem can be avoided by proper nailing of the roof members to the walls or by utilizing tie-down straps.

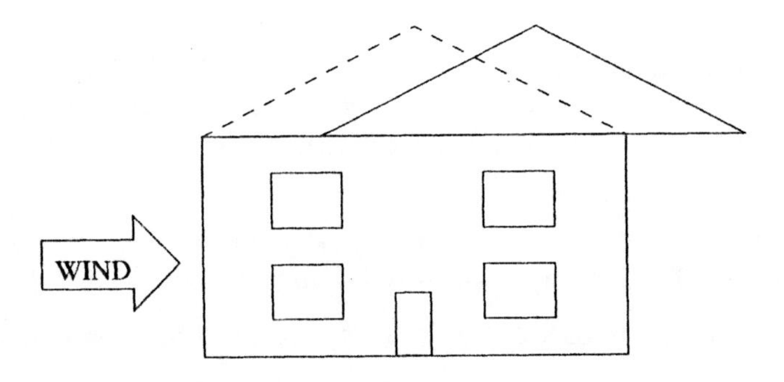

Figure 9-3—Failure of connection of roof to walls

Figure 9-4 depicts a failure that is a result of the roof being pulled up off the wall. The force that causes this failure is the suction force that is produced by wind forces going over the structure. The roof must be securely anchored to the walls to prevent this uplift.

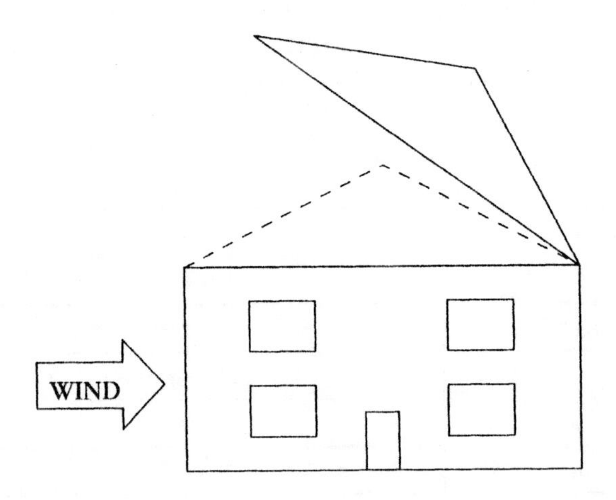

Figure 9-4—Separation of roof from wall by suction force

Advanced Discussion

The governing building code defines the wind loads that are to be applied to the structure. Once these loads are determined, a structural analysis can be performed to evaluate the stresses in the structure.

The wind pressure that will be experienced by the structure is determined by utilizing the following mathematical expression:

$$\text{Wind force} = 0.00256 \times \text{velocity}^2 \times \text{ modification factors}$$

The first two terms in this expression represent an equation commonly used in physics that the mass of a material times its velocity squared is equal to the force. The building code prescribes the appropriate velocity to place into the equation. The velocity chosen is basically a wind speed that will occur only once every 50 years. Values of the required wind velocities for selected location are given in Table 9-1. For specific locations, the local building code should be consulted.

Table 9-1—Maximum Wind Velocity Used in Wind Force Equation

Location	Range of Wind Velocity of One Occurrence in 50 Years (mph)
Alaska	70–110
Arkansas	70
Florida	80–110
Georgia	70–100
Iowa	70–80
Kentucky	70
Louisiana	70–100
Nebraska	80
North Dakota	70–90

The force that is determined using the first two terms of the equation is not adequate for design. Modifications are necessary to adjust the forces to account for different terrains, different plan dimensions of buildings, height of the building, etc. The modification factors are from ASCE7-1993 are:

I = Importance factor
C_e = Height, exposure and gust factor
C_p = Pressure coefficient

Importance Factor

The importance factor provides for an increase in the wind force for essential structures. Essential structures include fire stations, police stations, hospitals and other similar facilities. The theory behind this increase in design forces is the necessity of these structures remaining operational in the event of a disaster.

In contrast, the importance factor does allow for a reduction in wind forces for structures that would represent a low hazard to human life if they failed. These types of structures include temporary facilities, agricultural facilities, and minor storage buildings.

Table 9-2 shows the importance factors for various categories of structures.

Table 9-2—Importance Factors

Category	Description	Factor
I	All buildings not in categories II, III, and IV	1.0
II	Buildings that hold 300 or more people in one area	Increase by 15% (1.15)
III	Essential facilities	Increase by 15% (1.15)
IV	Low hazard buildings	Decrease by 10% (0.90)

Height, Exposure and Gust Factor

The modification factor, C_e, takes into account that different height structures of different shapes in different terrains will experience different wind forces.

Because the nature of the surroundings have an impact on the magnitude of the force, it is necessary to divide terrains into different categories, called *exposures*, as follows:

Exposure A = Center of large cities.

Exposure B = Suburban areas.

Exposure C = Flat and open country and grasslands.

Exposure D = Flat coastal areas subject to wind blowing over bodies of water.

Each of these categories has corresponding modification factors that take into account changes in the wind force from the exposure type and turbulence (i.e., gusts). For Exposure A, the modification factors for structures up to 30 ft in height are given in Table 9-3.

Table 9-3—C_e for Exposure A

Height above Ground (ft)	C_e
0–15	0.28
20	0.33
25	0.36
30	0.38

Table 9-3 provides the values for Exposure A. As a rough rule of thumb, the values of C_e for residential structures in this height range in Exposures B, C, and D are the values of C_e listed in Table 3 multiplied by 2, 3.5 and 4.5, respectively.

Pressure Coefficient

The pressure coefficient accounts for the shape of the structure in determining the wind force. For walls, this modification factor is 1.3 for buildings that are fairly square in plan.

The value of C_p for a roof depends on the height and pitch of the roof. For a roof with a 30-degree angle from the horizontal degree, a value of 0.9 providing uplift would be required.

Sample Calculations

Determine the wind force at the top of a 15-ft wall under the listed conditions.

Case 1: Kentucky: $V = 70$ mph (103 ft/sec.)

Low-hazard structure: $I = 0.90$

Center of large city (Exp. A): $C_e = 0.28$

Nearly square in plan $C_p = 1.3$

$$\text{Wind pressure} = 0.00256 \times V^2 \times I \times C_e \times C_p$$
$$= 0.00256 \times (103)^2 \times 0.90 \times 0.28 \times 1.3$$
$$= 8.9 \text{ psf}$$

Case 2: Northeastern Iowa: $V = 90$ mph (132 ft/sec.)

Essential facility: $I = 1.15$

Open Country (Exp. C): $C_e = 0.28 \times 3.5 = 0.98$

Nearly square in plan: $C_p = 1.3$

$$\text{Wind pressure} = 0.00256 \times V^2 \times I \times C_e \times C_p$$
$$= 0.00256 \times (132)^2 \times 1.15 \times 0.98 \times 1.3$$
$$= 65.4 \text{ psf}$$

In summary, note the following:

- Wind forces on one face of the structure can produce suction forces on the opposite face of the building.
- Horizontal wind forces cause uplift forces on portions of the roof.

- Wind force failures can occur if the building is not anchored properly to the foundation or if the roof is not properly tied to the walls.
- The wind force is a function of the velocity of the wind. The building codes require the wind force to be modified.
- Modification factors take into account the importance of the structure, height, terrain and wind gusts.

Analysis

Once the loads are known, the structure is analyzed to determine the impact of these forces. The analysis of most structures is complicated, often requiring the use of computers. For the most part, the analysis of residential structures can be accomplished with straight-forward computations.

The analysis portion of this text is presented in the following two chapters. The first chapter assists in the determination of how much deflection will result under the applied loads. The second chapter assists in the determination of the forces resulting when loads are applied.

10

Forces

Introduction

An adequately designed beam must satisfy both serviceability and strength requirements. Serviceability requirements are restrictions that ensure that the beam will perform in a manner that will not be annoying to its users nor cause damage to other components of the structure. Serviceability requirements are for the most part satisfied by limiting the amount of deflection of the beam. Other examples of serviceability requirements include setting minimum dimensions for the beams or providing a minimal acceptable limits of material properties.

Strength requirements of members are not related to performance characteristics but rather are related to the stress in the material. That is, the beam must be sufficiently designed such that it will not fail.

To avoid failure or distress, the beam must be able to support the applied loads. These loads will produce forces at the beam supports, shear forces and bending moments.

The loads imposed on the beam are transferred to the walls or columns that support it. The loads that the beam transfer to the columns or walls are called *support forces*. A beam supported by underdesigned columns or walls is of little value. To design the column or wall, the support forces must be known.

The applied loads will induce a shear force in the beam. A shear force will cause one section of the beam to slip past another section of the beam. This is shown in Figure 10-1.

A beam will deflect downward or sometimes upward when a load is applied. This deflection will increase or decrease along the length of the member, giving it a curved shape. This curved shape, or rotation, will produce stresses that are both compressive and tensile. This is because one side of the beam is experiencing shortening (compressive) while the other side is elongating (tensile). The rotational force causing this stress is called the *bending moment.*

Thus the support reaction is needed to design the supporting column. The shear force and bending moment are needed to design the beam. This chapter provides the maximum values for these three items for a variety of loading and beam conditions. At the end of this chapter illustrative examples are provided.

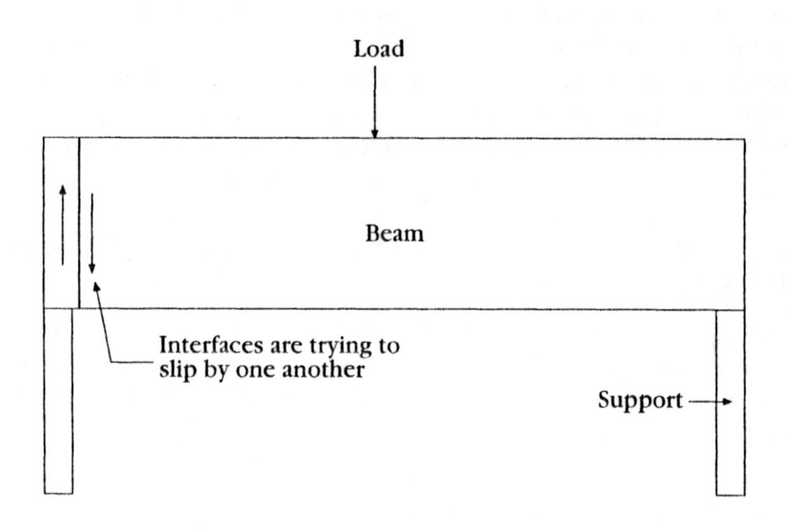

Figure 10-1—Shearing forces in a beam

Case 1

Description: Uniformly distributed load on entire length of simply supported beam.

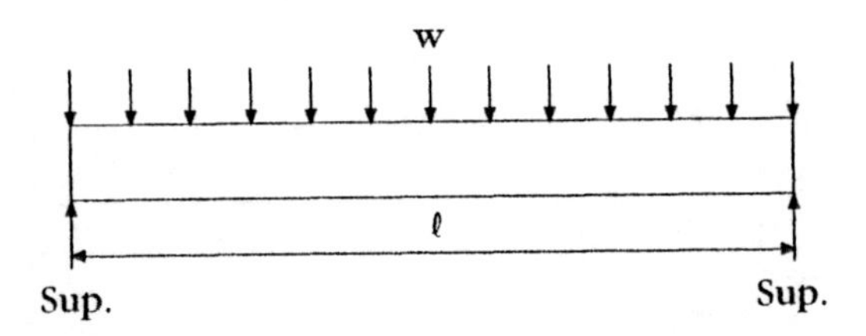

Maximum bending moment = M = $w\ell^2/8$ (Center)

Maximum shear = V = $w\ell/2$ end

Force on left support = R_L = $w\ell/2$

Force on right support = R_R = $w\ell/2$

Case 2

Description: Uniformly distributed load on half of simply supported beam.

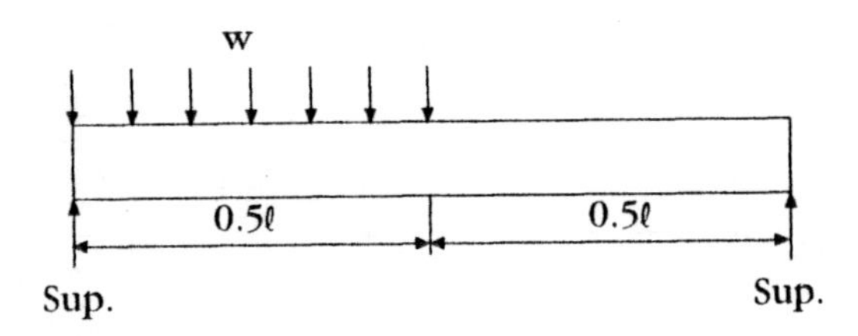

Maximum bending moment = M = $9w\ell^2/128$ (Near center)

Maximum shear = V = $3w\ell/8$ (Left end)

Force on left support = R_L = $3w\ell/8$

Force on right support = R_R = $w\ell/8$

Case 3

Description: Uniformly distributed load on one-quarter of simply supported beam.

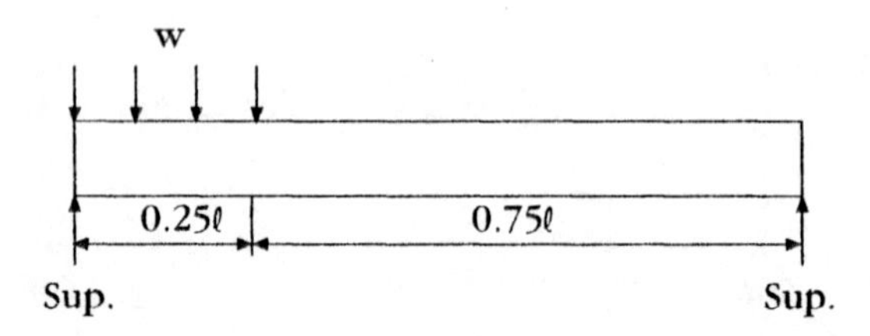

Maximum bending moment = M = $49w\ell^2/2048$ (Near quarter point)

Maximum shear = V = $7w\ell/32$ (Left end)

Force on left support = $R_L = 7w\ell/32$

Force on right support = $R_R = w\ell/32$

Case 4

Description: Concentrated load at midspan of simply supported beam.

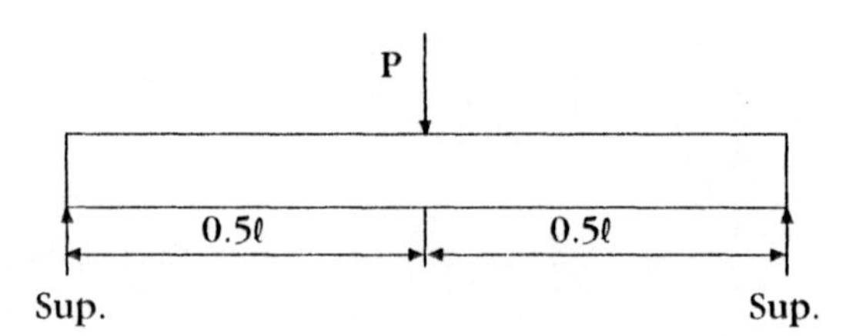

Maximum bending moment = M = $P\ell/4$ (Center)

Maximum shear = V = $P/2$ (Ends)

Force on left support = $R_L = P/2$

Force on right support = $R_R = P/2$

Case 5

Description: Concentrated load located at 0.4ℓ from end of simply supported beam.

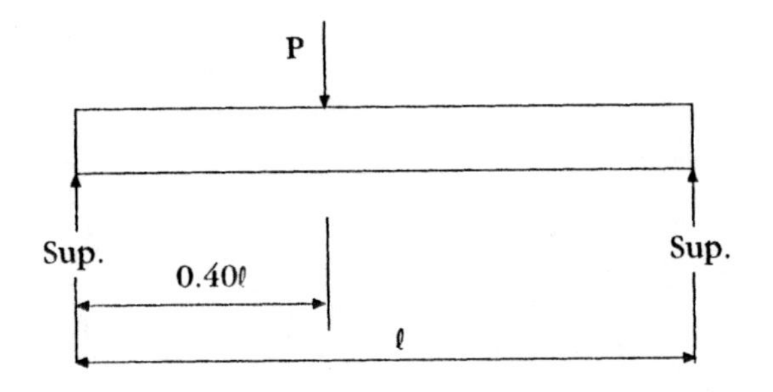

Maximum bending moment = M = $6P\ell/25$ (At load)

Maximum shear = V = 3P/5 (Left side of beam)

Force on left support = R_L = 3P/5

Force on right support = R_R = 2P/5

Case 6

Description: Concentrated load located at 0.3ℓ from end of simply supported beam.

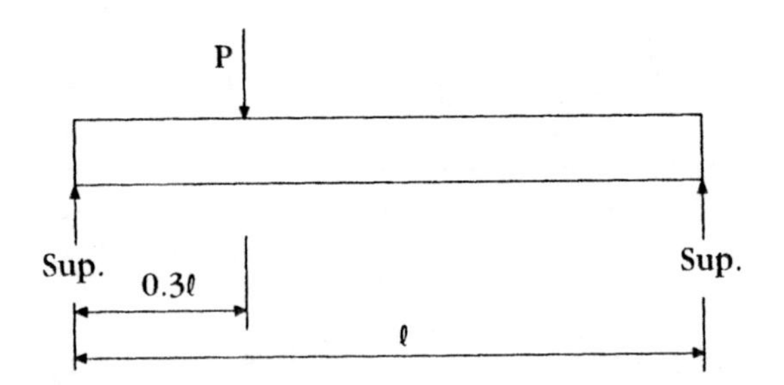

Maximum bending moment = M = $21P\ell/100$ (At load)

Maximum shear = V = 7P/10 (Left end)

Force on left support = 7P/10

Force on right support = 3P/10

Case 7

Description: Concentrated load located at 0.2ℓ from end of simply supported beam.

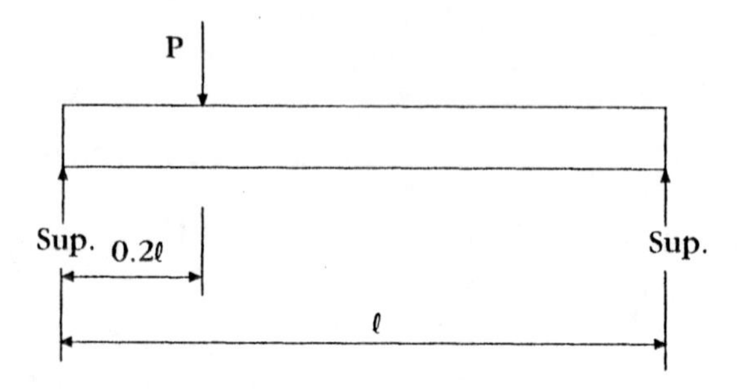

Maximum bending moment = M = $4P\ell/25$

Maximum shear = V = $4P/5$ (Left end)

Force on left support = $4P/5$

Force on right end = $P/5$

Case 8

Description: Uniformly distributed load across entire cantilever beam.

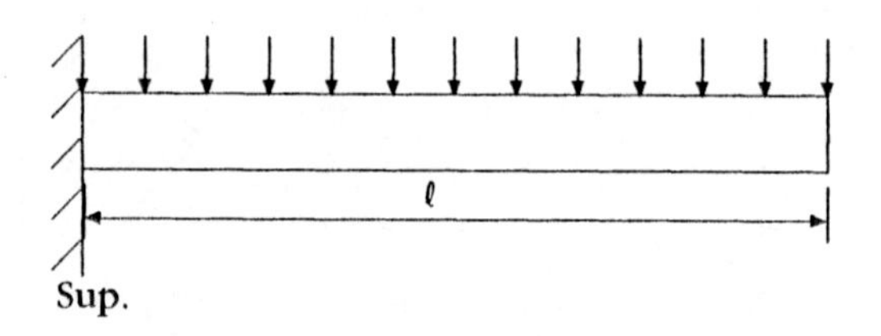

Maximum bending moment = $w\ell^2/2$ (At support)

Maximum shear = V = $w\ell$ (At support)

Force on support = R = $w\ell$

Case 9

Description: Uniformly distributed load on left half of a cantilever beam.

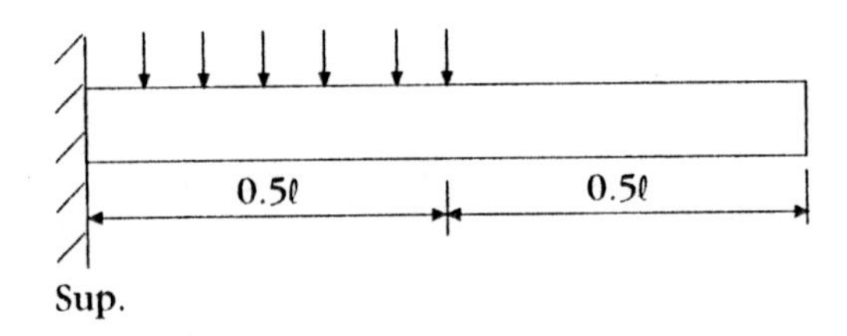

Maximum bending moment = M = $w\ell^2/8$ (At support)

Maximum shear = V = $w\ell/2$ (At support)

Force on support = R = $w\ell/2$

Case 10

Description: Uniformly distributed load on left one-quarter of cantilever beam.

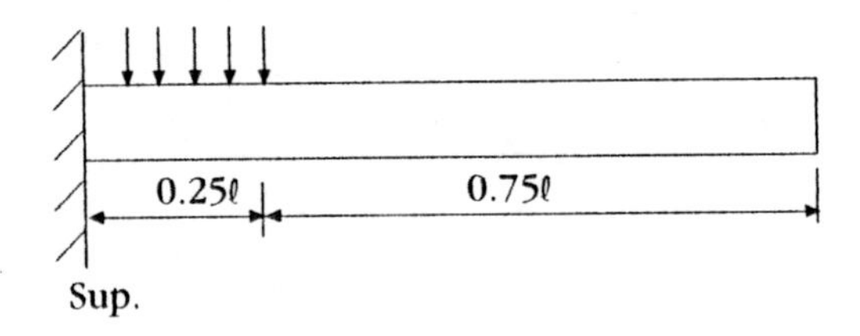

Maximum bending moment = M = $w\ell^2/32$ (At support)

Maximum shear = V = $w\ell/4$ (At support)

Force on support = R = $w\ell/4$

Case 11

Description: Concentrated load on the end of a cantilever beam.

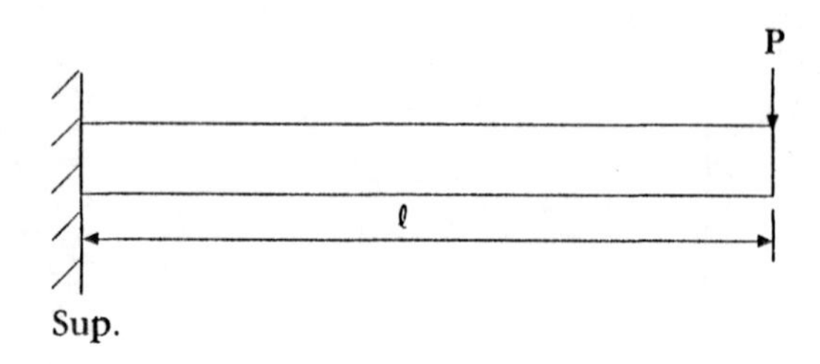

Maximum bending moment = M = Pℓ (At support)

Maximum shear = V = P (At support)

Force on support = R = P

Case 12

Description: Concentrated load at the quarter-point from the right end.

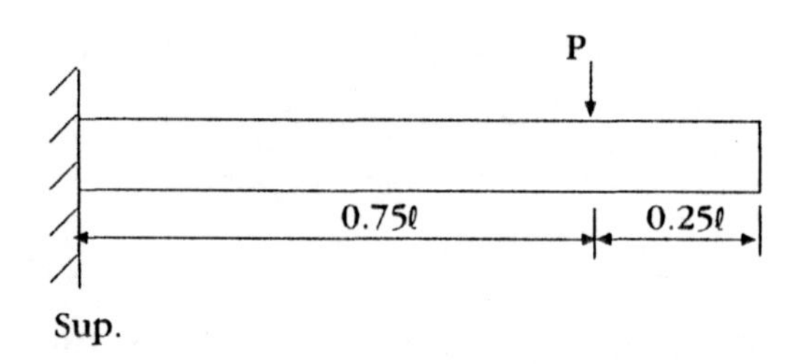

Maximum bending moment = M = 3Pℓ/4 (At support)

Maximum shear = V = P (At support)

Force at support = R = P

Case 13

Description: Concentrated load at the midpoint of a cantilever beam.

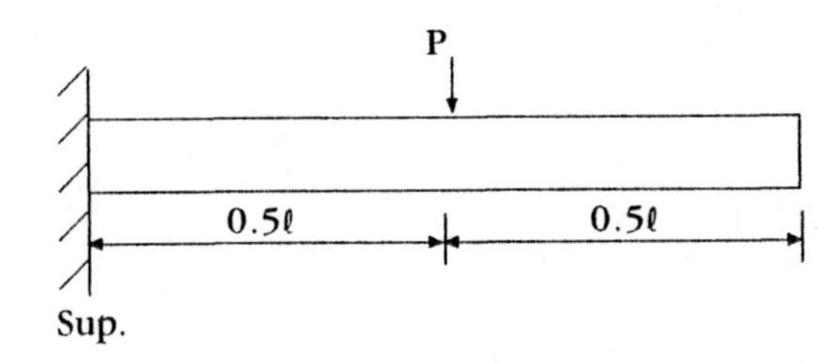

Maximum bending moment = M = Pℓ/2 (At support)

Maximum shear = V = P (At support)

Force at support = P

Case 14

Description: Concentrated load at the quarter point from the left end.

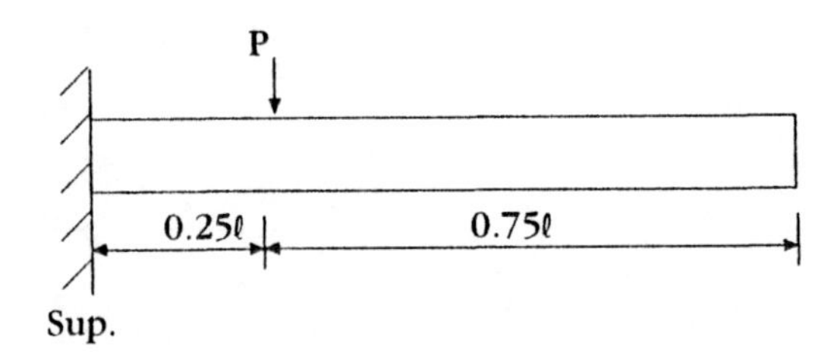

Maximum bending moment = M = Pℓ/4 (At support)

Maximum shear = V = P (At support)

Force at support = P

Case 15

Description: Beam with concentrated load on overhang that is 20% of beam length.

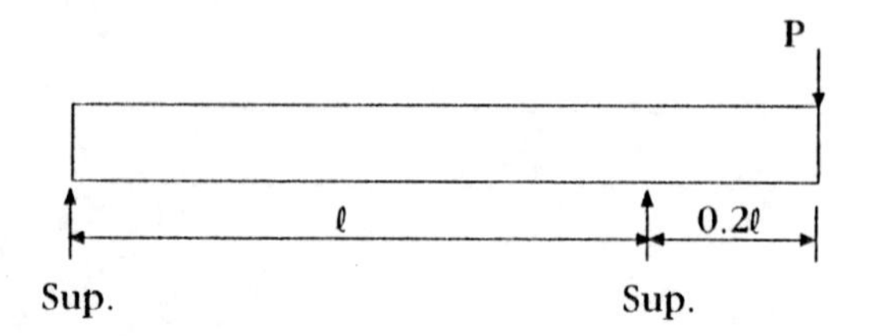

Maximum bending moment = M = $P\ell/5$ (At right support)

Maximum shear = V = P (In overhang)

Force on left support = R_L = P/5 (Up)

Force on right support = R_R = 6P/5 (Down)

Case 16

Description: Beam with concentrated load on overhang that is 30% of beam length.

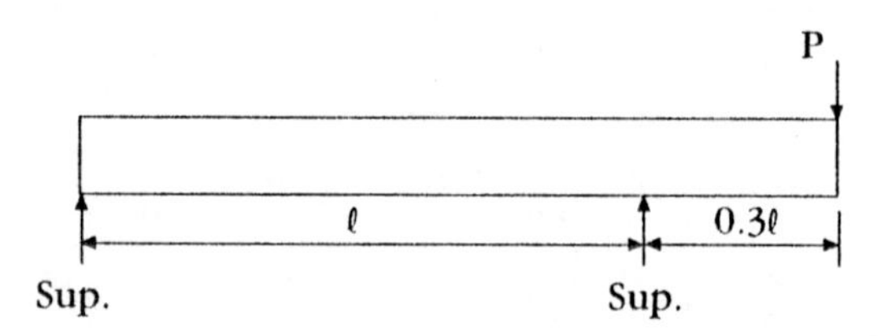

Maximum bending moment = M = $3P\ell/10$ (At right support)

Maximum shear = V = P (In overhang)

Force on left support = R_L = 3P/10 (Up)

Force on right support = R_R = 13P/10 (Down)

Case 17

Description: Beam with concentrated load on overhang that is 40% of beam length.

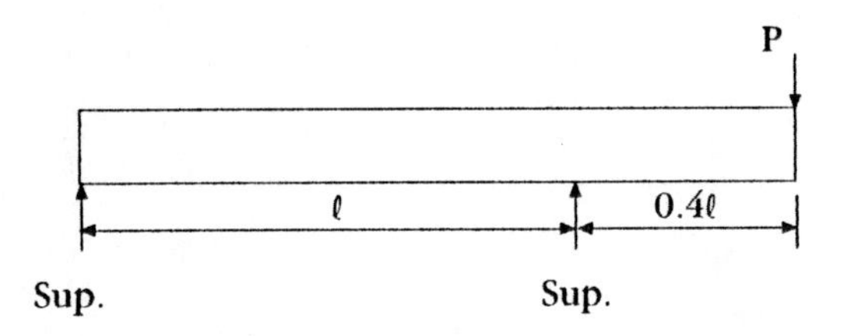

Maximum bending moment = M = $2P\ell/5$ (At right support)

Maximum shear = V = P (In overhang)

Force on left support = $2P/5$ (Up)

Force on right support = $7P/5$ (Down)

Case 18

Description: Uniformly distributed load on overhang of beam. Overhang length is 30% of beam length.

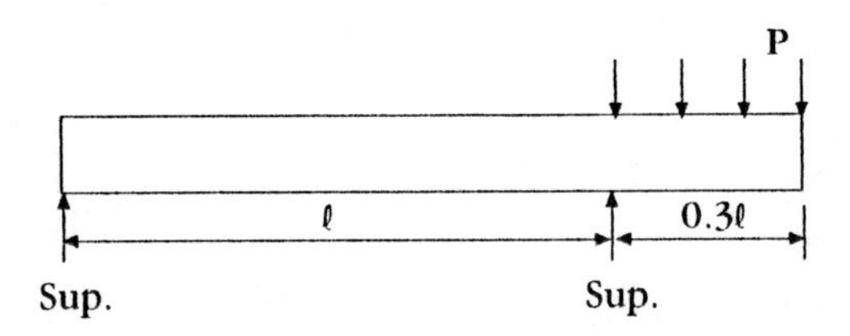

Maximum bending moment = M = $9w\ell^2/200$ (At right support)

Maximum shear = V = $3w\ell/10$ (At right support)

Force on left support = $R_L = 9w\ell/200$ (Up)

Force on right support = $R_R = 69w\ell/200$ (Down)

Case 19

Description: Uniform load over entire length of beam with overhang. Overhang length is 30% of beam length.

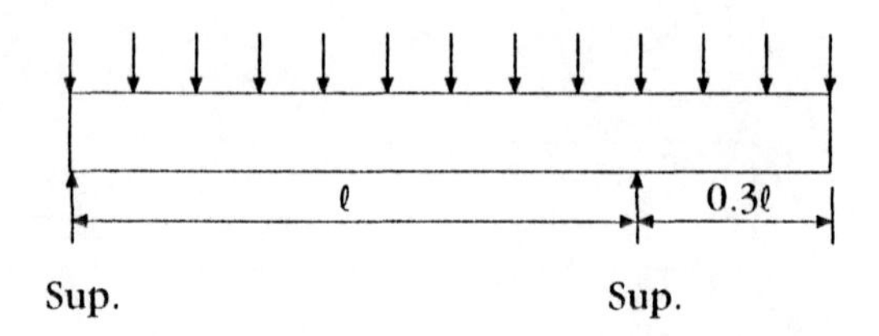

Maximum bending moment = M = $w\ell^2/10$ (Near midspan)

Maximum shear = V = $109w\ell/200$ (At right support)

Force at left support = R_L = $91w\ell/200$

Force at right support = R_R = $169w\ell/200$

Case 20

Description: Uniformly distributed load in span for beam with overhang. Overhang length is 30% of beam length.

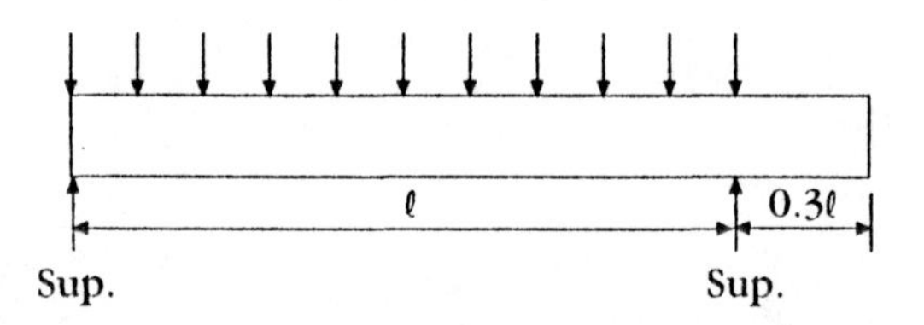

Maximum bending moment = M = $w\ell^2/8$ (Center)

Maximum shear = V = $w\ell/2$ (At support)

Force on left support = R_L = $w\ell/2$

Force on right support = R_R = $w\ell/2$

Case 21

Description: Triangular load on simply supported beam.

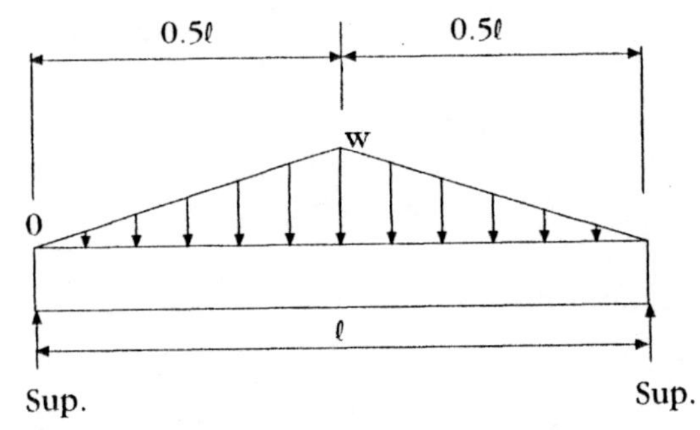

Maximum bending moment = M = $w\ell^2/12$ (At center)

Maximum shear = V = $w\ell/4$ (At support)

Force on each support = R = $w\ell/4$

Example 1

This example will compare the results of a load distributed over the entire span to that of a concentrated load.

Determine the maximum bending moment, shear force and support for the following beam under the given conditions.

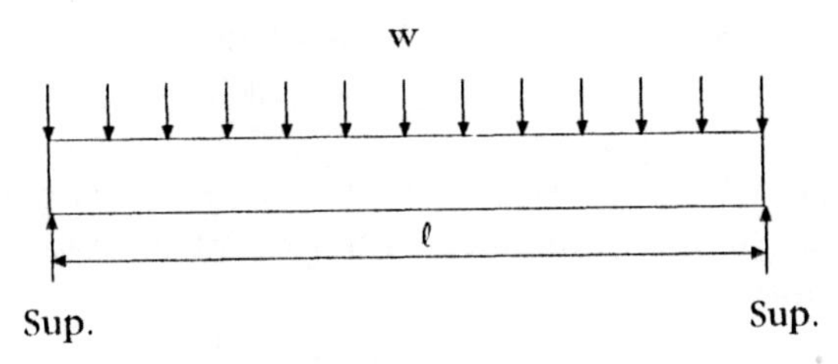

w = Distributed load on beam
= 60 lb/linear ft
= 5 lb/linear in.

ℓ = Length of beam
= 12 ft
= 144 in.

Referring to the information provided for Case 1:

$$\text{Maximum bending moment} = M = w\ell^2/8 = (5)(144)^2/8 = 12960 \text{ in.-lb}$$

$$\text{Maximum shear} = V = w\ell/2 = (5)(144)/2 = 360 \text{ lb}$$

$$\text{Maximum support force} = R = w\ell/2 = 360 \text{ lb}$$

Compare these values to the results if the distributed load is lumped into one concentrated load at midspan. The concentrated load is equal to the distributed load multiplied by the beam length:

$$P = w\ell = (5)(144) = 720 \text{ lb}$$

Referring to the information provided for Case 4, the maximum results are:

$$\text{Maximum bending moment} = M = P\ell/4 = (720)(144)/4 = 25920 \text{ in.-lb}$$

$$\text{Maximum shear} = V = P/2 = 720/2 = 360 \text{ lb}$$

$$\text{Maximum support force} = P/2 = 360 \text{ lb}$$

Comparison of the results show that the shear and support force remain unchanged but the heavy load at midspan doubled the bending moment.

Example 2

This example compares the results of a load distributed over the entire span to that of a concentrated load.

Determine the maximum bending moment, shear force and support force for the following cantilever beam under the given conditions.

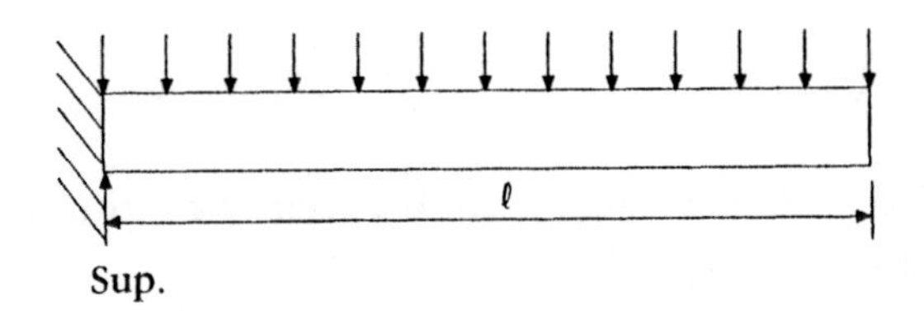

w = Distributed load
= 60 lb/linear ft
= 5 lb/linear in.

ℓ = Length of beam
= 4 ft
= 48 in.

Referring to the information provided for Case 8:

$$\text{Maximum bending moment} = M = w\ell^2/2 = (5)(48)^2/2 = 5760 \text{ in.-lb}$$

$$\text{Maximum shear} = V = w\ell = (5)(48) = 240 \text{ lb}$$

$$\text{Maximum support force} = R = w\ell = 240 \text{ lb}$$

Compare these values to the results if the distributed load is lumped into one concentrated load at midspan. The concentrated load is equal to the distributed load multiplied by the beam length:

$$P = w\ell = (5)(48) = 240 \text{ lb}$$

Referring to the information provided for Case 13, the maximum results are:

Maximum bending moment = M = $P\ell/2$
= (240)(48)/2
= 5760 in.-lb

Maximum shear = V = P = 240 lb

Maximum support force = R = P = 240 lb

Comparison of the results shows that they produce identical maximums.

Advanced Discussion

The 21 load cases that have been presented thus far in this chapter all had two supports per beam, except the cantilevers. If a beam has more than two supports, it is called a *continuous beam*, as opposed to the two-support condition, which is called a *simple span*. The results for a simple-span beam are significantly different than the same-length beam with three supports. This section presents results for the following cases:

Case 22: Two-span continuous beam with one span loaded. (Three supports)

Case 23: Two-span continuous beam with both spans loaded. (Three supports)

Case 24: Three-span continuous beam with two exterior spans loaded. (Four supports)

Case 25: Three-span continuous beam with three adjacent spans loaded. (Four supports)

Case 22

Description: Uniform load on one span of a two-span beam.

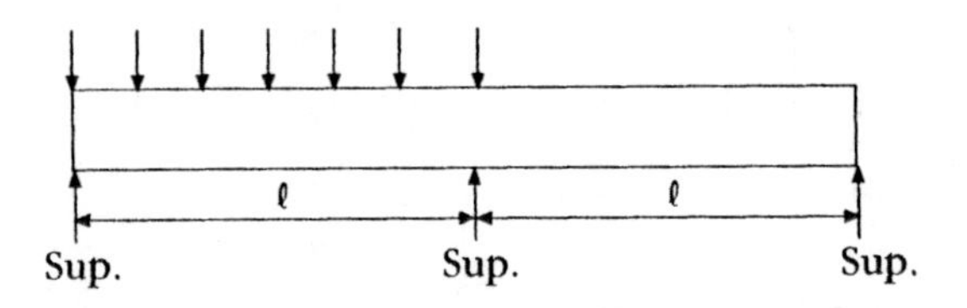

Maximum bending moment = M = $49w\ell^2/512$ (Near center of loaded span)

Maximum shear = V = $9w\ell/16$ (At center support)

Force on left support = $R_L = 7w\ell/16$

Force on center support = $R_C = 5w\ell/8$

Force on right support = $R_R = w\ell/16$ (up)

Case 23

Description: Uniform load on both spans of a two-span beam.

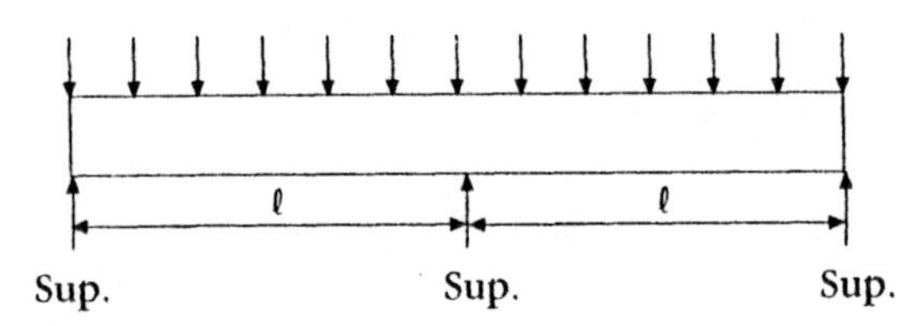

Maximum bending moment = M = $w\ell^2/14$ (Near center of span)

Maximum shear = V = $5w\ell/8$ (At center support)

Force on end supports = $R_L = R_R = 3w\ell/8$

Force on center support = $R_C = 5w\ell/4$

Case 24

Description: Uniform load on the end span of a three-span beam.

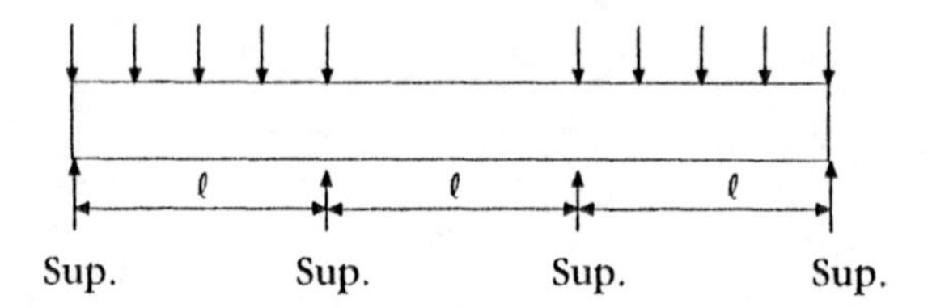

Maximum bending moment = $M = w\ell^2/10$ (Near center of end span)

Maximum shear = $V = 5w\ell/9$ (At center supports)

Force on end supports = $R_E = 5w\ell/11$

Force on interior supports = $R_I = 5w\ell/9$

Case 25

Description: Uniform load on all spans of a three-span beam.

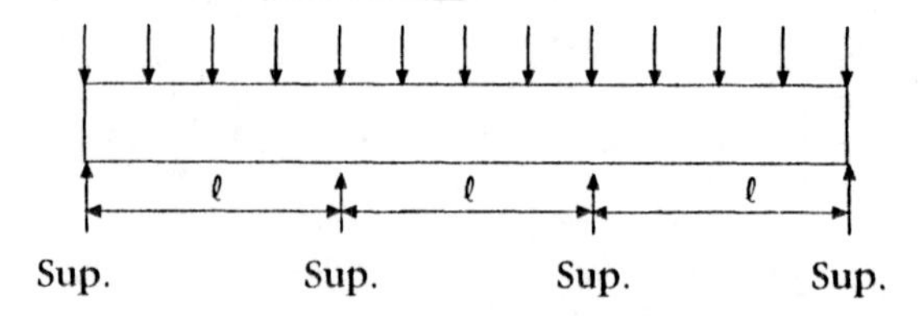

Maximum bending moment = $M = w\ell^2/10$ (Reverse moment at interior support)

Maximum shear = $V = 3w\ell/5$ (At interior support)

Force on end support = $R_e = 2w\ell/5$

Force on interior support = $R_I = 11w\ell/10$

Example 3

Compare the interior support force for a two-span and three-span beam as shown below. Assume the distributed load is the same for both cases.

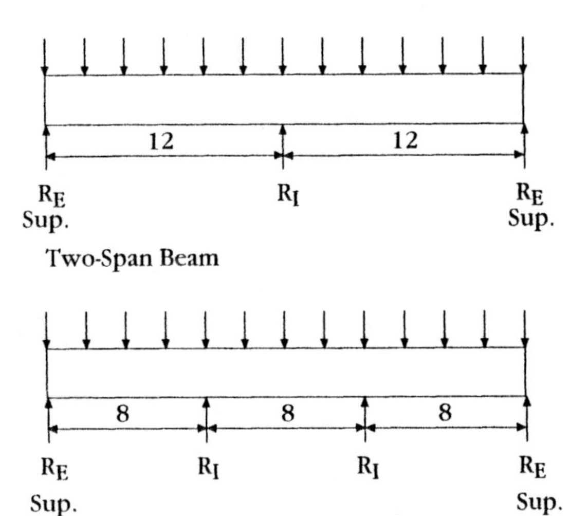

Three-Span Beam

For the two-span beam the interior support force is given by (see Case 23):

$$
\begin{aligned}
R_I &= 5w\ell/4 \\
&= 5w\,(12 \times 12)/4 \\
&= 180w
\end{aligned}
$$

For the three-span beam the interior support force is given by (see Case 25):

$$
\begin{aligned}
R_I &= 11w\ell/10 \\
&= 11w\,(8 \times 12)/10 \\
&= 106w
\end{aligned}
$$

Therefore, by the addition of an extra support, the interior support force is reduced by:

$$\frac{106}{180} = 0.59$$

$$
\begin{aligned}
1 - 0.59 &= 0.41 \\
&= 41\%
\end{aligned}
$$

11

Deflections

Introduction

A structural member must have sufficient strength to carry the applied loads with tolerable deflections. Even if a member has sufficient strength it will not be acceptable if there is unsightly deflections in the span. These undesirable deflections will not only look unsightly but they might result in cracking in masonry, plaster and architectural finishes.

Various codes and specifications provide guidelines for the maximum allowable deflections. The following are two of these guidelines.

Floor live load deflection = span length/360

Beams supporting masonry = span length/600

Note that deflection limits are the same regardless of the supporting material used.

Example: A 12-ft long beam is used to support the floor live loads. What is the maximum allowable deflection?

For a 12-ft long beam, the maximum allowable deflection is the span length divided by 360.

$$\text{Maximum deflection} = (12 \times 12)/360 = 0.4 \text{ in.}$$

The introduction portion of this chapter presents the deflections for 21 different beam conditions. The advanced discussion presents four additional cases.

Introduction

The following beam conditions are provided in this section:

Case 1: Simply Supported-Uniform Load

Case 2: Simply Supported-Half Beam Loaded

Case 3: Simply Supported-Quarter Beam Loaded

Case 4: Simply Supported-Midspan Load

Case 5: Simply Supported-Load at 0.4L

Case 6: Simply Supported-Load at 0.3L

Case 7: Simply Supported-Load at 0.2L

Case 8: Cantilever-Uniform Load

Case 9: Cantilever-Half Beam Loaded

Case 10: Cantilever-Quarter Beam Loaded

Case 11: Cantilever-End Load

Case 12: Cantilever- Load at 3/4L

Case 13: Cantilever-Load at Midspan

Case 14: Cantilever-Load at 1/4L

Case 15: Overhang (20%)-Load on Overhang

Case 16: Overhang (30%)-Load on Overhang

Case 17: Overhang (40%)-Load on Overhang

Case 18: Overhang (30%)-Distributed Load on Overhang

Case 19: Overhang (30%)-Full Distributed Load

Case 20: Overhang (30%)-Distributed Load in Span

Case 21: Triangular Load

Case 1

Description: Uniformly distributed load on entire length of simply supported beam.

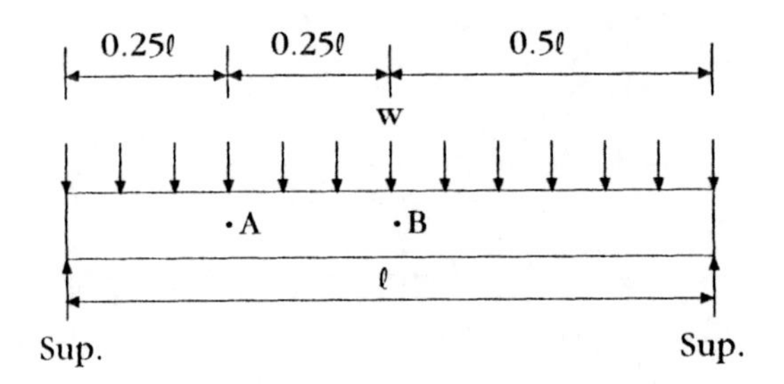

Deflection at A = $w\ell^4/(108EI)$ 71% of maximum

Deflection at B = $w\ell^4/(77EI)$ maximum

Example

Calculate the deflection at midspan for the following conditions:

w = Load on beam
= 55 lb/linear ft
= 4.6 lb/linear in.

ℓ = Length of beam
= 12 ft
= 144 in.

E = Modulus of elasticity
= 1,400,000 lb/in.^{2}

I = Moment of inertia
= 178 in.^{4}

Δ_B = $w\ell^4/(108EI)$
= $(4.6)(144)^4/(77 \times 1{,}400{,}000 \times 178)$
= 0.103 in.
= Approx. 3/32 in.

Case 2

Description: Uniformly distributed load on half of simply supported beam.

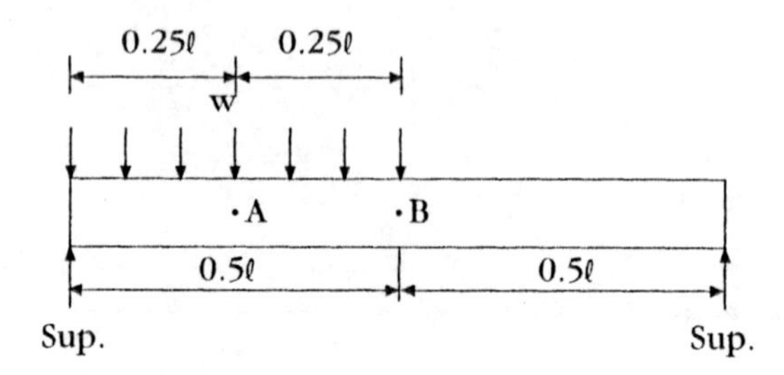

Deflection at A = $w\ell^4/(198EI)$ 77% of maximum

Deflection at B = $w\ell^4/(154EI)$ 99% of maximum

Example

Calculate the maximum deflection for the following conditions:

w = Load on beam
= 55 lb/linear ft
= 4.6 lb/linear in.

ℓ = Length of beam
= 12 ft
= 144 in.

E = Modulus of elasticity
= 1,400,000 lb/in.

I = Moment of Inertia
= 178 in.^{4}

Δ_B = $w\ell^4/(154EI)$
= $4.6(144)^4/(154 \times 1{,}400{,}000 \times 178)$
= 0.0515 in.

Δ_B = 99% of Δ_{max}

Therefore: Δ_{max} = $\Delta_B/0.99$
= 0.052 in.
= Approx. 3/64 in.

Case 3

Description: Uniformly distributed load on one-quarter of simply supported beam.

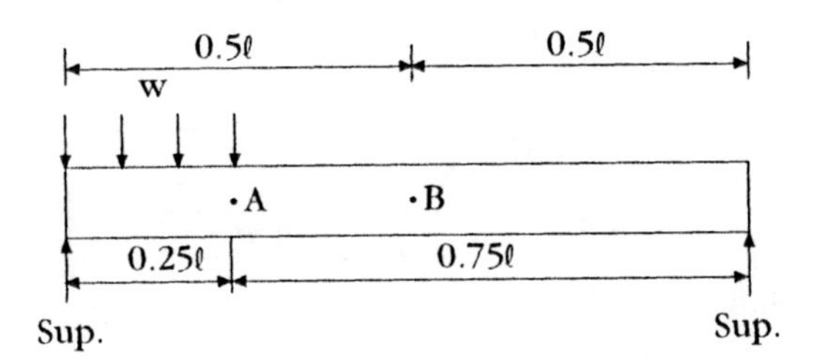

Deflection at A = $w\ell^4/(630EI)$ 83% of maximum

Deflection at B = $w\ell^4(534EI)$ 98% of maximum

Example

Calculate the deflection at midspan for the following conditions:

w = Load on beam
= 55 lb/linear ft
= 4.6 lb/linear in.

ℓ = Length of beam
= 12 ft
= 144 in.

E = Modulus of elasticity
= 1,400,000 lb/in.

I = Moment of inertia
= 178 in.^{4}

Δ_B = $w\ell^4/(535EI)$
= $(4.6)(144)^4/(535 \times 1{,}400{,}000 \times 178)$
= 0.015 in.
= Approx 1/64 in.

Case 4

Description: Concentrated load at midspan of simply supported beam.

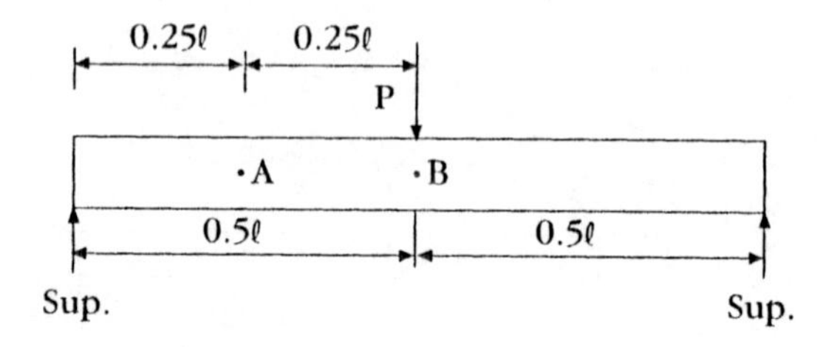

Deflection at A = $P\ell^3/(70EI)$ 69% of maximum

Deflection at B = $P\ell^3/(48EI)$ Maximum

Example

Calculate the deflection at midspan for the following conditions:

P = Concentrated load
= 10,000 lb

ℓ = Length of beam
= 12 ft
= 144 in.

E = Modulus of elasticity
= 29,000,000 lb/in.^{2}

I = Moment of inertia
= 48 in.^{4}

Δ_A = $P\ell/(48EI)$
= $(10{,}000)(144)^3/(48 \times 29{,}000{,}000 \times 48)$
= 0.45 in.
= Approx 1/2 in.

Case 5

Description: Concentrated load located at 0.4ℓ from end of simply supported beam.

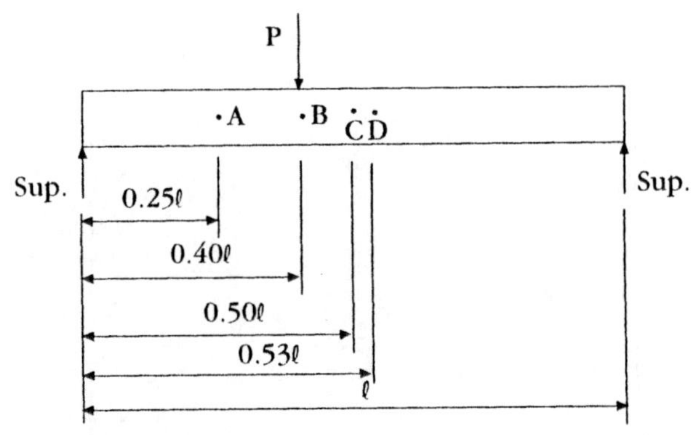

Deflection at A = $P\ell^3/(69EI)$	73% of maximum
Deflection at B = $P\ell^3/(52EI)$	97% of maximum
Deflection at C = $P\ell^3/(51EI)$	99.6% of maximum
Deflection at D = $P\ell^3/(51EI)$	Maximum

Example

Calculate the deflection at the quarter point near the left end of the beam for the following conditions:

P = Concentrated load
= 10,000 lb

ℓ = Length of beam
= 12 ft
= 144 in.

E = Modulus of elasticity
= 29,000,000 lb/in.^{2}

I = Moment of inertia
= 48 in.^{4}

Δ_A = $P\ell^3/(69EI)$
= $(10{,}000)(144)^3/(69 \times 29{,}000{,}000 \times 48)$
= 0.31
= Approx. 5/16 in.

Case 6

Description: Concentrated load located at 0.3ℓ from end of simply supported beam.

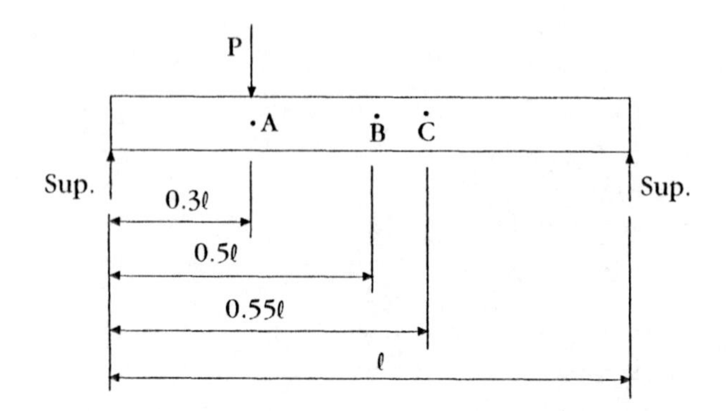

Deflection at A = $P\ell^3/(68EI)$	88% of maximum
Deflection at B = $P\ell^3/(61EI)$	99% of maximum
Deflection at C = $P\ell^3/(60EI)$	Maximum

Example

Calculate deflection at midspan for following conditions:

P = Concentrated load
= 10,000 lb

ℓ = Length of beam
= 12 ft
= 144 in.

E = Modulus of elasticity
= 29,000,000 lb/in.^{2}

I = Moment of inertia
= 48 in.^{4}

Δ_B = $P\ell^3/(61EI)$
= $(10{,}000)(144)^3/(61 \times 29{,}000{,}000 \times 48)$
= 0.35 in.
= Approx. 11/32 in.

Case 7

Description: Concentrated load located at 0.2ℓ from end of simply supported beam.

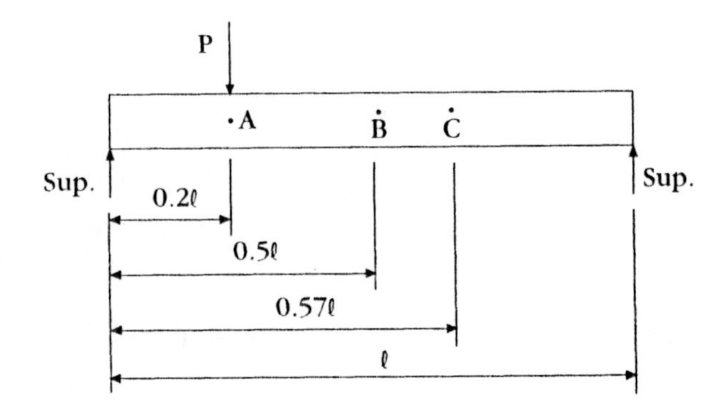

Deflection at A = $P\ell^3/(117EI)$	71% of maximum
Deflection at B = $P\ell^3/(85EI)$	98% of maximum
Deflection at C = $P\ell^3/(83EI)$	Maximum

Example

Calculate deflection at the location of the concentrated load for the following conditions:

P = Concentrated load
= 10,000 lb

ℓ = Length of beam
= 12 ft
= 144 in.

E = Modulus of elasticity
= 29,000,000 lb/in.^{2}

I = Moment of inertia
= 48 in.^{4}

Δ_A = $P\ell^3/(117EI)$
= $(10{,}000)(144)^3/(117 \times 29{,}000{,}000 \times 48)$
= 0.18 in.
= Approx. 3/16 in.

Case 8

Description: Uniformly distributed load across entire cantilever beam.

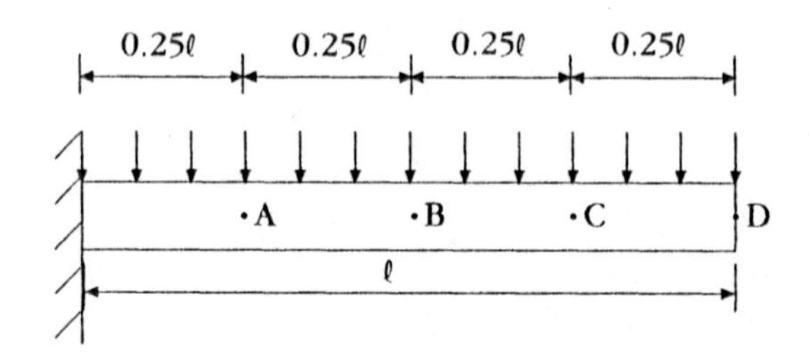

Deflection at A = $w\ell^4/(76EI)$ 11% of maximum

Deflection at B = $w\ell^4/(23EI)$ 35% of maximum

Deflection at C = $w\ell^4/(12EI)$ 67% of maximum

Deflection at D = $w\ell^4/(8EI)$ Maximum

Example

Calculate the deflection at the free end of the beam for the following conditions:

w = Distributed load on beam
= 500 lb/linear ft
= 41.7 lb/linear in.

ℓ = Length of cantilever beam
= 4 ft
= 48 in.

E = Modulus of elasticity
= 1,400,000 lb/in.^{2}

I = Moment of inertia
= 178 in.^{4}

Δ_D = $w\ell^4/(8EI)$
= $(41.7)(48)^4/(8 \times 1{,}400{,}000 \times 178)$
= 0.11 in.
= Approx. 1/8 in.

Case 9

Description: Uniformly distributed load on left half of a cantilever beam.

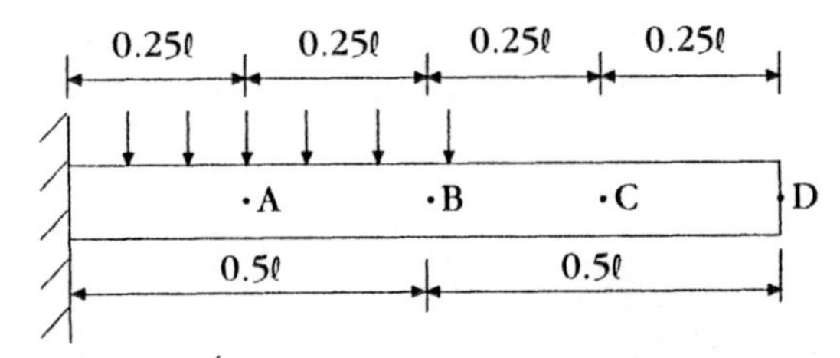

Deflection at A = $w\ell^4/(361EI)$	15% of maximum
Deflection at B = $w\ell^4/(128EI)$	43% of maximum
Deflection at C = $w\ell^4/(77EI)$	71% of maximum
Deflection at D = $w\ell^4/(55EI)$	Maximum

Example

Calculate the additional deflection that occurs after the load has ended for the following conditions:

w = Distributed load on beam
= 500 lb/linear ft
= 41.7 lb/linear in.

ℓ = Length of cantilever beam
= 4 ft
= 48 in.

E = Modulus of elasticity
= 1,400,000 lb/in.^{2}

I = Moment of inertia
= 178 in.^{4}

This calculation is performed by determining the difference between the deflection at the end and the deflection at midspan.

Δ_D $= w\ell^4/(55EI)$
$= (41.7)(48)^4/(55 \times 1{,}400{,}000 \times 178)$
$= 0.016$

Δ_B $= w\ell^4/(128EI)$
$= (41.7)(48)^4/(128 \times 1{,}400{,}000 \times 178)$
$= 0.007$

$\Delta_D - \Delta_B$ $= 0.016 - 0.007$
$= 0.009$ in.

Approximately 0.009 (approx. 1/100 in.) of additional deflection occurs after the point the load has terminated.

Case 10

Description: Uniformly distributed load on left one-quarter of cantilever beam.

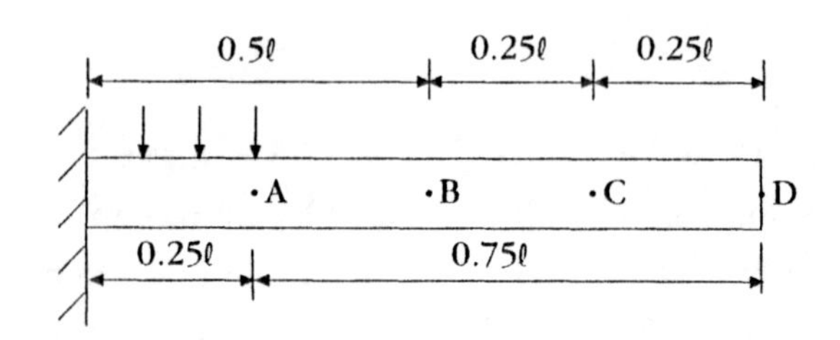

Deflection at A = $w\ell^4/(2048EI)$	20% of maximum
Deflection at B = $w\ell^4/(878EI)$	47% of maximum
Deflection at C = $w\ell^4/(559EI)$	73% of maximum
Deflection at D = $w\ell^4/(410EI)$	Maximum

Example

Calculate the deflection at the location where the distributed load ends for the following condition:

w = Distributed load on beam
= 500 lb/linear ft
= 41.7 lb/linear in.

ℓ = Length of cantilever beam
= 4 ft
= 48 in.

E = Modulus of elasticity
= 1,400,000 lb/in.^{2}

I = Moment of inertia
= 178 in.^{4}

Δ_A = $w\ell^4/(2048EI)$
= $(41.7)(48)^4/(2048 \times 1{,}400{,}000 \times 178)$

= 0.0004 in.
= Approx. 1/2000 in.

Case 11

Description: Concentrated load on end of a cantilever beam.

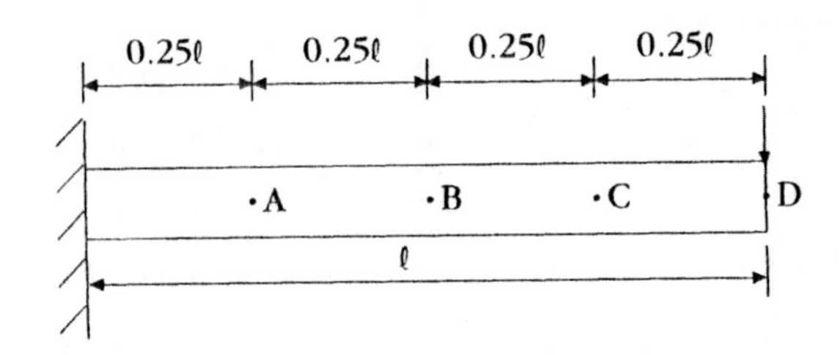

Deflection at A = $P\ell^3/(35EI)$ 9% of maximum

Deflection at B = $P\ell^3/(10EI)$ 31% of maximum

Deflection at C = $P\ell^3/(5EI)$ 63% of maximum

Deflection at D = $P\ell^3/(3EI)$ Maximum

Example

Calculate the maximum deflection for the following conditions:

P = Concentrated load
= 1,000 lb

ℓ = Length of beam
= 4 ft
= 48 in.

E = Modulus of elasticity
= 1,400,000 lb/in.^{2}

I = Moment of inertia
= 48 in.^{4}

Δ_D = $P\ell^3/(3EI)$
= $(1000)(48)^3/(3 \times 1{,}400{,}000 \times 48)$
= 0.55 in.
= Approx. 1/2 in.

Case 12

Description: Concentrated load at quarter point from the right end.

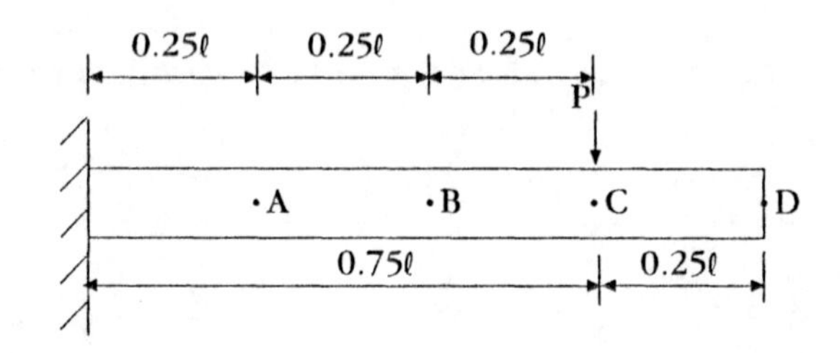

Deflection at A = $P\ell^3/(48EI)$	10% of maximum
Deflection at B = $P\ell^3/(14EI)$	35% of maximum
Deflection at C = $P\ell^3/(7EI)$	67% of maximum
Deflection at D = $P\ell^3/(5EI)$	Maximum

Example

Calculate the maximum deflection for the following conditions:

p = Concentrated load
= 1000 lb

ℓ = Length of beam
= 4 ft
= 48 in.

E = Modulus of elasticity
= 1,400,000 lb/in.^{2}

I = Moment of inertia
= 48 in.^{4}

Δ_D = $P\ell^3/(5EI)$
= $(1{,}000)(48)^3/(5 \times 1{,}400{,}000 \times 48)$
= 0.33 in.
= Approx. 5/16 in.

Case 13

Description: Concentrated load at midpoint of a cantilever beam.

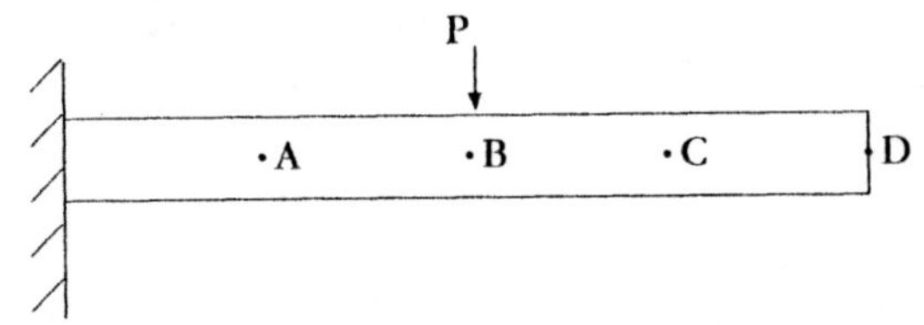

Deflection at A = $P\ell^3/(77EI)$ 13% of maximum

Deflection at B = $P\ell^3/(24EI)$ 42% of maximum

Deflection at C = $P\ell^3/(14EI)$ 71% of maximum

Deflection at D = $P\ell^3/(10EI)$ Maximum

Example

Calculate the load at midspan needed to cause a 1-in. deflection at the free end for the following conditions.

ℓ = Length of beam
= 6 ft
= 72 in.

E = Modulus of elasticity
= 1,400,000 lb/in.^{2}

I = Moment of inertia
= 48 in.^{4}

The deflection at the end of the beam is given by the following equation:

$$\Delta_D = P\ell^3/(10EI)$$

Including the information that is known into the equation gives:

$$1 = P(72)^3/(10 \times 1{,}400{,}000 \times 48)$$

$$1 = \frac{373248\ P}{672{,}000{,}000}$$

$$1 = 0.000555P$$

$$P = 1/0.000555$$

$$P = 1800 \text{ lb}$$

Case 14

Description: Concentrated load at quarter point from the left end.

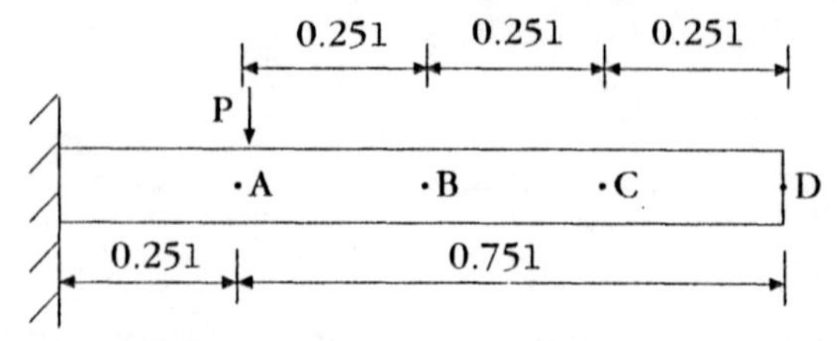

Deflection at A = $P\ell^3/(192EI)$ 18% of maximum

Deflection at B = $P\ell^3/(77EI)$ 45% of maximum

Deflection at C = $P\ell^3/(48EI)$ 73% of maximum

Deflection at D = $P\ell^3/(35EI)$ Maximum

Example

Calculate the maximum load that can be applied at Point A to cause a 1-in. deflection at the free end for the following conditions:

ℓ = Length of beam
= 6 ft
= 72 in.

E = Modulus of elasticity
= 1,400,000 lb/in.^{2}

I = Moment of inertia
= 48 in.^{4}

The deflection at the end of the beam is given by the following equation:

$$\Delta_D = P\ell^3/(35EI)$$

Substituting the information that is known into the equation gives:

$$1 \text{ in.} = P(72)^3/(35 \times 1{,}400{,}000 \times 48)$$

$$1 \text{ in.} = \frac{373248\ P}{2{,}352{,}000{,}000{,}000}$$

$$1 \text{ in.} = 0.00016P$$

$$P = 6301 \text{ lb}$$

Case 15

Description: Beam with concentrated load on overhang that is 20% of beam length.

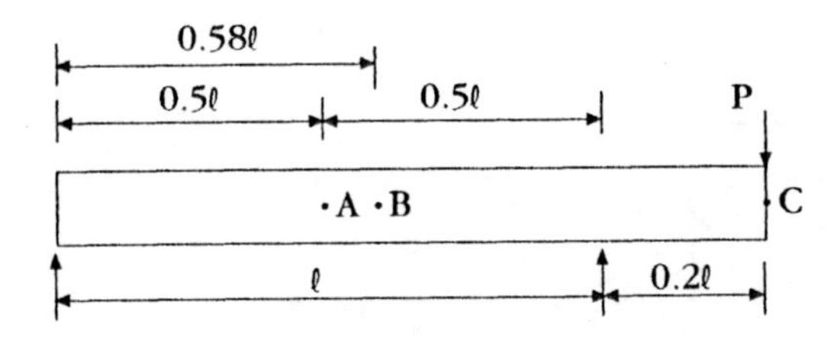

Upward deflection at A = $P\ell^3/(80EI)$

Upward deflection at B(0.58ℓ) = $P\ell^3/(78EI)$ Maximum in span

Downward deflection at C = $P\ell^3/(63EI)$ Maximum in overhang

Example

Calculate the distance that the center of the beam moves upward under the following conditions:

P = Concentrated load
= 10,000 lb

ℓ = Length of beam
= 10 ft
= 120 in.

Overhang = 2 ft
= 24 in.

E = Modulus of elasticity
= 1,400,000 lb/in.^{2}

I = Moment of inertia
= 178 in.^{4}

Δ_A = $P\ell^3/(80EI)$
= $(10{,}000)(120)^3/(80 \times 1{,}400{,}000 \times 178)$
= 0.87 in.
= 7/8 in.

Case 16

Description: Beam with concentrated load on overhang that is 30% of beam length.

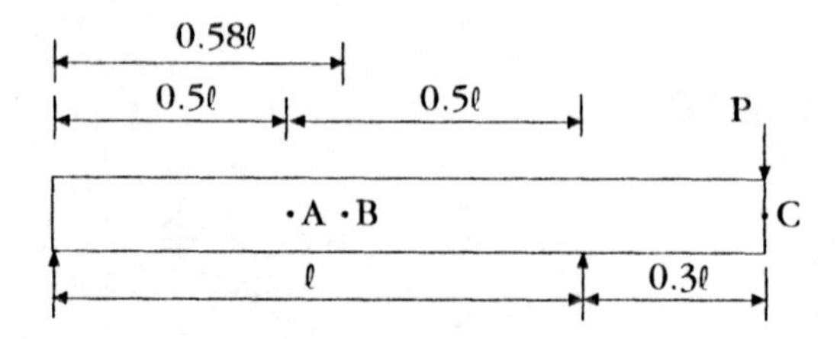

Upward deflection at A = $P\ell^3/(53EI)$

Upward deflection at B(0.58ℓ) = $P\ell^3/(52EI)$ Maximum in span

Downward deflection at C = $P\ell^3/(26EI)$ Maximum in overhang

Example

Calculate the maximum downward deflection in the beam under the following condition:

P = Concentrated load
= 10,000 lb

ℓ = Length of beam
= 10 ft
= 120 in.

Overhang = 3 ft
= 36 in.

E = Modulus of elasticity
= 1,400,000 lb/in.^{2}

I = Moment of inertia
= 178 in.^{4}

Δ_C = $P\ell^3/(26EI)$
= $(10{,}000)(120)^3/(26 \times 1{,}400{,}000 \times 178)$
= 2.67 in.
= 2-5/8 in.

Case 17

Description: Beam with concentrated load on overhang that is 40% of beam length.

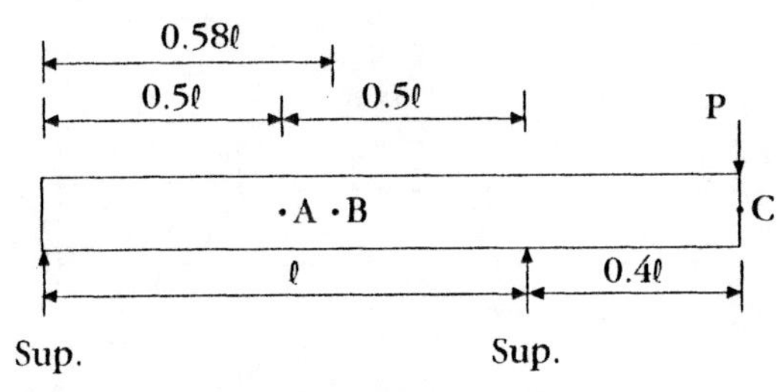

Upward deflection at A = $P\ell^3/(40EI)$

Upward deflection at B(0.58ℓ) = $P\ell^3/(39EI)$ — Maximum in span

Downward deflection at C = $P\ell^3/(13EI)$ — Maximum in overhang

Example

Calculate the maximum amount the beam will rise under the following conditions:

P = Concentrated load
= 10,000 lb

ℓ = Length of beam
= 10 ft
= 120 in.

Overhang = 4 ft
= 48 in.

E = Modulus of elasticity
= 1,400,000 lb/in.^{2}

I = Moment of inertia
= 178 in.^{4}

Δ_C = $P\ell^3/(39EI)$
= $(10{,}000)(120)^3/(39 \times 1{,}400{,}000 \times 178)$
= 1.78 in.
= 1-3/4 in.

Case 18

Description: Uniformly distributed load on overhang of beam. Overhang length is 30% of beam length.

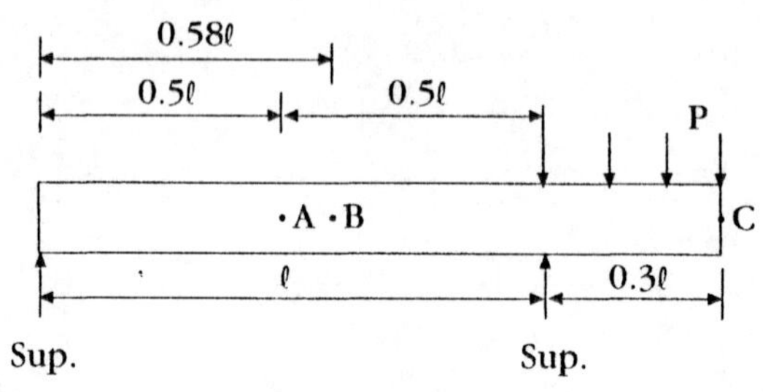

Upward deflection A = $w\ell^4/(356EI)$

Upward deflection B(0.58ℓ) = $w\ell^4/(346EI)$ Maximum in span

Downward deflection C = $w\ell^4/(181EI)$ Maximum

Example

Calculate the maximum downward deflection in the beam under the following conditions:

w = Uniformly distributed load
= 3340 lb/linear ft
= 278 lb/linear in.

ℓ = Length of beam
= 10 ft
= 120 in.

Overhang = 3 ft
= 36 in.

E = Modulus of elasticity
= 1,400,000 lb/in.^{2}

I = Moment of inertia
= 178 in.^{4}

Δ_B = $w\ell^4/(181EI)$
= $(278)(120)^4/(181 \times 1{,}400{,}000 \times 178)$
= 1.28 in.
= 1-1/4 in.

Case 19

Description: Uniformly load over entire length of beam with overhang. Overhang length is 30% of beam length.

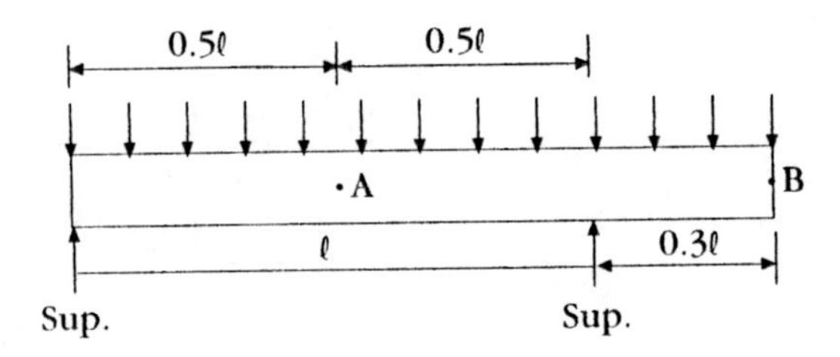

Downward deflection at A = $w\ell^4/(98EI)$

Upward deflection at B = $w\ell^4/(143EI)$

Example

Calculate the maximum upward deflection at B under the following conditions:

w = Uniformly distributed load
= 3340 lb/linear ft
= 278 lb/linear in.

ℓ = Length of beam
= 10 ft
= 120 in.

Overhang = 3 ft
= 36 in.

E = Modulus of elasticity
= 1,400,000 lb/in.^{2}

I = Moment of inertia
= 178 in.^{4}

Δ_B = $w\ell^4/(143EI)$
= $(278)(120)^4/(143 \times 1{,}400{,}000 \times 178)$
= 1.62 in.
= 1-5/8 in.

Case 20

Description: Uniformly distributed load in span for beam with overhang. Overhang length is 30% of beam length.

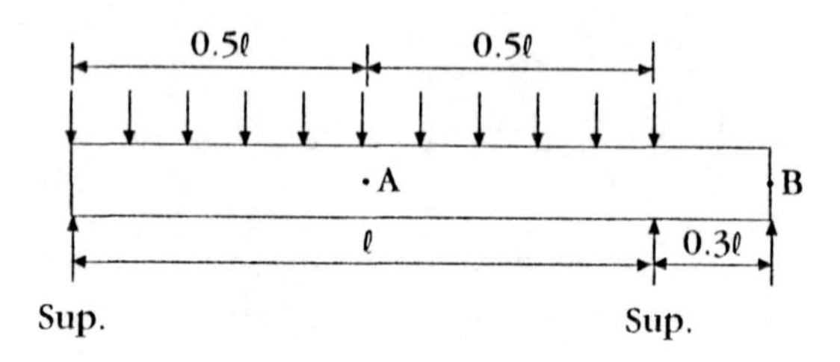

Downward deflection at A = $w\ell^4/(77EI)$

Upward deflection at B = $w\ell^4/(80EI)$

Example

Calculate the maximum upward deflection at B under the following conditions:

w = Uniformly distributed load
= 3340 lb/linear ft
= 278 lb/linear in.

ℓ = Length of beam
= 10 ft
= 120 in.

Overhang = 3 ft
= 36 in.

E = Modulus of elasticity
= 1,400,000 lb/in.^{2}

I = Moment of inertia
= 178 in.^{4}

Δ_B = $w\ell^4/(80EI)$
= $(278)(120)^4/(80 \times 1{,}400{,}000 \times 178)$
= 2.89 in.
= 2-7/8 in.

Case 21

Description: Triangular load on simply supported beams.

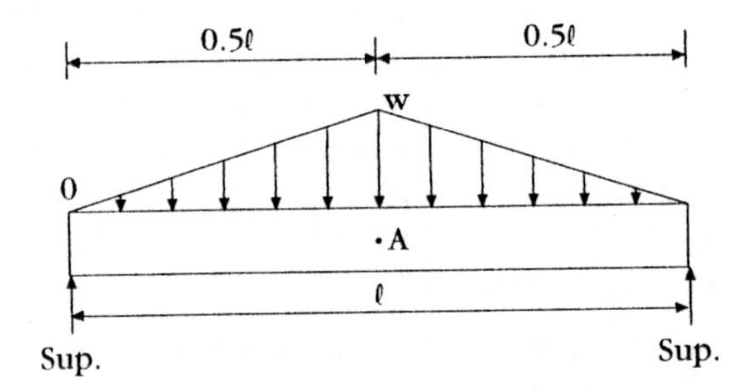

Deflection at A = $W\ell^3/(60EI)$

Example

Calculate the deflection at midspan for the following conditions:

W = Total load on beam
= $w\ell/2$

w = 660 lb/linear ft
= 55 lb/linear in.
= (55)(144)/2
= 3960 lb

ℓ = Length of beam
= 12 ft
= 144 in.

E = Modulus of elasticity
= 1,400,000 lb/in.^{2}

I = Moment of inertia
= 178 in.^{4}

Δ_A = $(3960)(144)^3/(60 \times 1{,}400{,}000 \times 178)$
= 0.79 in.
= Approx. 3/4 in.

Advanced Discussion

The 21 load cases that have been presented thus far in this chapter all had two supports per beam, except the cantilever beam. If a beam has more than two supports, it is called a *continuous beam*, as opposed to the two-support condition, which is called a *simple span*. The results for a simple span beam is significantly different than the same-length beam with three supports. This section will present results for the following cases:

Case 22: Two-span continuous beam with one span loaded. (Three supports)

Case 23: Two-span continuous beam with both spans loaded. (Three supports)

Case 24: Three-span continuous beam with two exterior spans loaded. (Four supports)

Case 25: Three-span continuous beam with three spans loaded. (Four supports)

Case 22

Description: Two-span continuous beam with uniform load in one span.

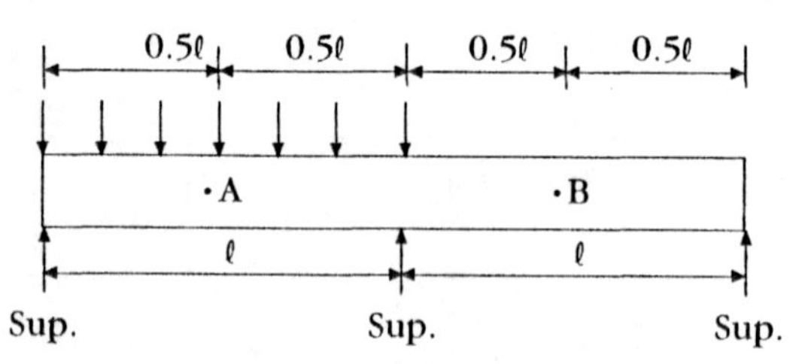

Downward deflection at A = $w\ell^4/(110EI)$

Upward deflection at B = $w\ell^4/(250EI)$

Example

The loading has caused the deflection at point A to be 1 in. What is the upward deflection at B, under the following conditions?

ℓ = Span length
= 6 ft
= 72 in.

E = Modulus of elasticity
= 1,400,000 lb/in.^{2}

I = Moment of inertia
= 178 in.^{4}

Δ_A = 1 in. = $w\ell^4/(110EI)$
1 in. = $(w)(72)^4/(110 \times 1{,}400{,}000 \times 178)$
1 in. = w/1020
w = 1020 lb/linear in.

The upward deflection at Point B with a load of 1020 lb/linear in. is:

Δ_B = $w\ell^4/(250EI)$
= $(1020)(72)^4/(250 \times 1{,}400{,}000 \times 178)$
= 0.44 in.
= Approx. 7/16 in. upward

Case 23

Description: Two-span continuous beam with uniform load in both spans.

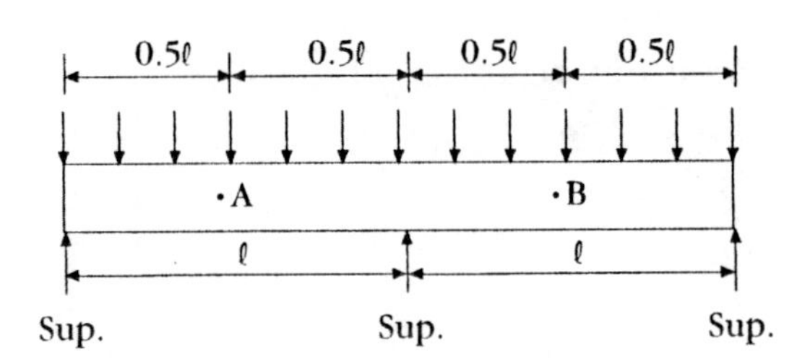

Deflection at A = $w\ell^4/(192EI)$

Deflection at B = Deflection at A

Example

Determine the uniform load needed to cause a 1-in. deflection at point A under the following conditions:

ℓ = Span length
= 6 ft
= 72 in.

E = Modulus of elasticity
= 1,400,000 lb/in.^{2}

I = Moment of inertia
= 178 in.^{4}

Δ_A = $1 = w\ell^4/(192EI)$
$1 = w(72)^4/(192 \times 1{,}400{,}000 \times 178)$
$1 = w/1780$
w = 1780 lb/linear in.
= 148 lb/linear ft

Case 24

Description: Three-span continuous beam with two spans loaded.

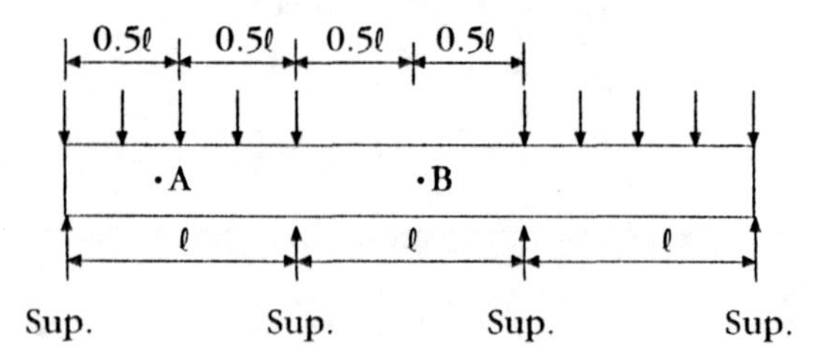

Downward deflection at A = $w\ell^4/(100EI)$

Upward deflection at B = $w\ell^4/(160EI)$

Example

What load will result in a 1/2-in. deflection at A under the following conditions:

ℓ = Span length
= 10 ft
= 120 in.

E = Modulus of elasticity
= 1,400,000 lb/in.^{2}

I = Moment of inertia
= 178 in.^{4}

Δ_A = $0.5 = w\ell^4/(100EI)$
$0.5 = w(120)^4/(100 \times 1,400,000 \times 178)$
$0.5 = w/120$
w = 60 lb/linear in.
= 5 lb/linear ft

Case 25

Description: Three-span continuous beam with all spans loaded.

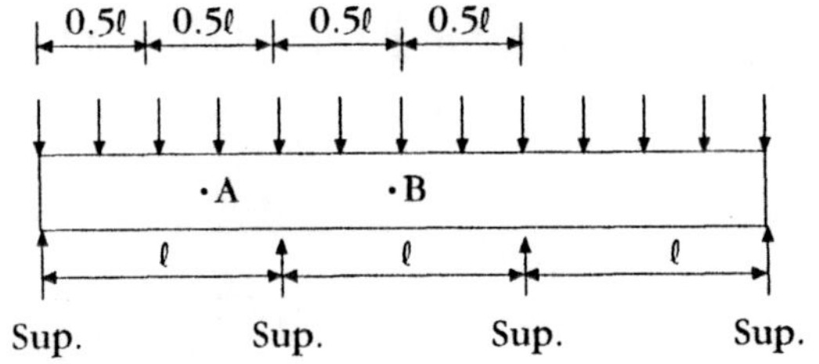

0.5ℓ 0.5ℓ 0.5ℓ 0.5ℓ
·A ·B
ℓ ℓ ℓ
Sup. Sup. Sup. Sup.

Deflection at A = $w\ell^4/(145EI)$

Deflection at B = $w\ell^4/(1927EI)$

Example

If the deflection at A is 1 in., what is the deflection at B?

$$\Delta_A = 1 \text{ in.} = w\ell^4/(145EI)$$
$$145 = w\ell^4/(EI)$$

$$\Delta_B = w\ell^4/(1927EI)$$
$$= 1/1927 \times w\ell^4/EI$$

Substitute 145 for $w\ell^4/(EI)$

$$\Delta_B = 1/1927 \times 145$$

$$\Delta_B = 145/1927$$
$$= 0.08 \text{ in.}$$

Deflections were presented for 21 different load cases. In the advanced discussion of this chapter, more complicated conditions are also presented. Note that all of these loading conditions can be used in combination with one another to determine deflections of many other loading conditions. Combining these load conditions is called *super-positioning*. The theory of super-positioning provides that the deflection caused by several loads is equal to the deflection determined by summing the deflection of each load separately. This theory is explained in the following example.

Determine the deflection at midspan for the beam shown with the given conditions.

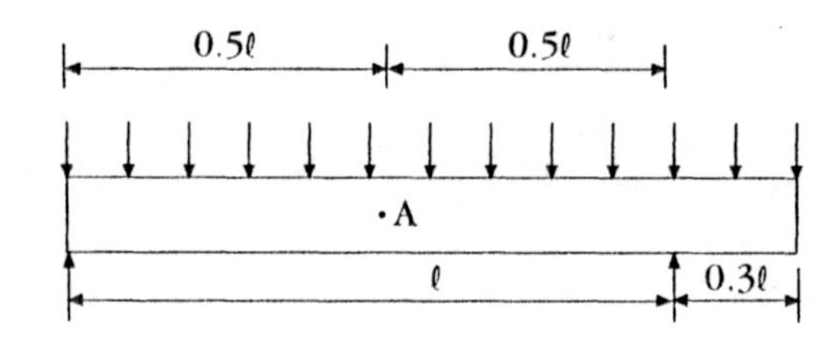

w = Distributed load
= 1400 lb/linear ft
= 117 lb/linear in.

ℓ = Length of beam
= 10 ft
= 120 in.

Overhang length = 3 ft
= 36 in.

E = Modulus of elasticity
= 1,400,000 lb/in.^{2}

I = Moment of inertia
= 178 in.^{4}

With the information provided in Case 19 utilized, the deflection at midspan is:

$\Delta = w\ell^4/(98EI)$
$= (117)(120)^4/(98 \times 1{,}400{,}000 \times 178)$
$= 1.0$ in.

This same result can be obtained by combining the information provided in Cases 18 and 20. The deflection determined using Case 18 is:

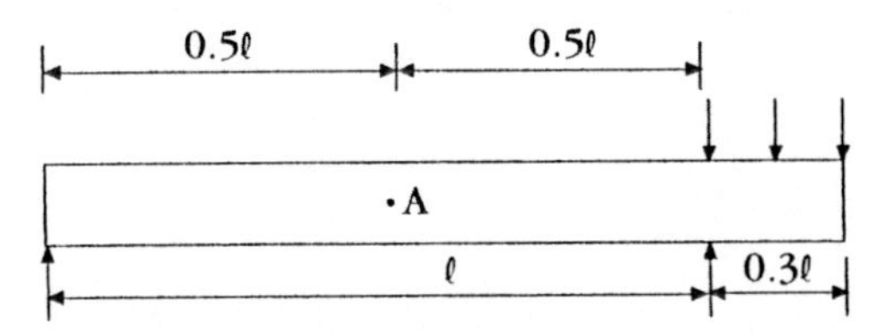

Upward deflection at A $= w\ell^4/(356EI)$
$= (117)(120)^4/(356 \times 1{,}400{,}000 \times 178)$
$= 0.3$ in. upward

The deflection determined using Case 20 is:

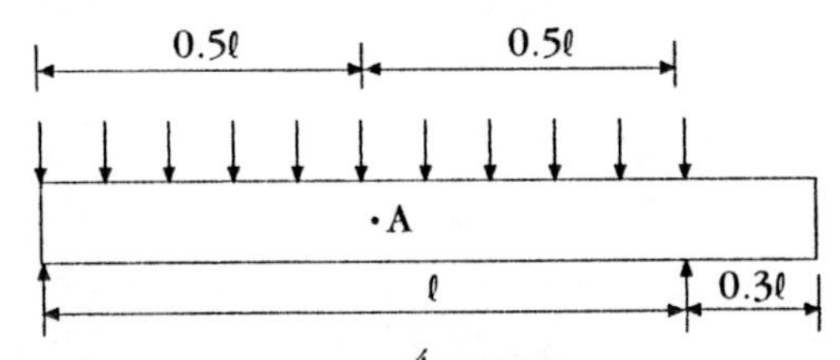

Downward deflection at A $= w\ell^4/(77EI)$
$= (117)(120)^4/(77 \times 1{,}400{,}000 \times 178)$
$= 1.3$ in. down

Adding these two deflection (1.3 + (-0.3)) = 1.0 in. The following provides a summary of the previous calculations.

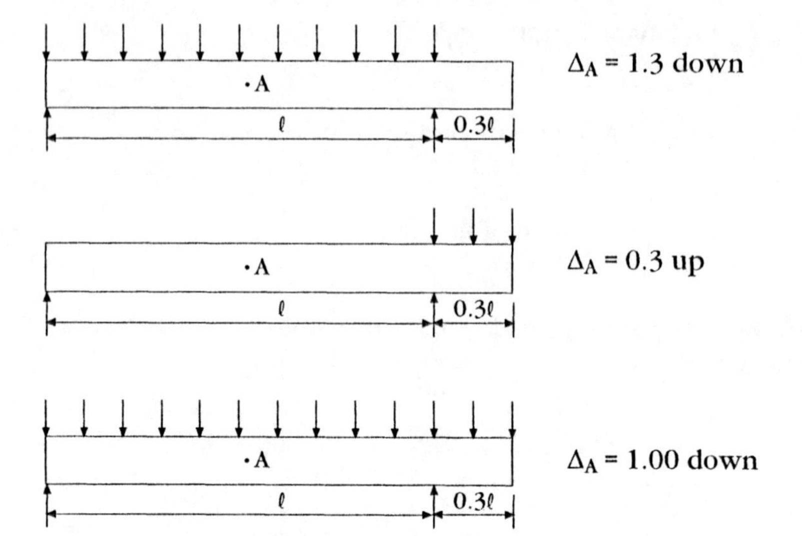
A
ℓ
0.3ℓ
ΔA = 1.3 down
A
ℓ
0.3ℓ
ΔA = 0.3 up
A
ℓ
0.3ℓ
ΔA = 1.00 down

Wood Members

Wood has been the traditional building material for residential construction because of its abundance and because it is easy to work with. Home construction in the United States has grown from the log cabin to the conventional framing used in modern times. The discussion on wood members in the following three chapters includes detailed explanations on two types of beams. The first, joists, are closely spaced members that carry the floor and roof loads. The other chapter on beams covers the design of heavily loaded beams used over door and window openings. These beams are called *headers*.

The other chapter on wood members is on trusses. Trusses are commonly used to carry the roof loads. The truss chapter presents the different trusses used in the residential construction market.

12

Wood Joists

Introduction

The loads on the floors and the ceilings are carried by joists. Joists are shallow, closely spaced structural members that are made of wood, steel, concrete or even a combination of these materials. Concrete and steel joists are presently limited to commercial and industrial construction. When utilized in residential construction, joists are exclusively wood.

Figures 12-1 and 12-2 show views of a wood joist system. The joists may be as shallow as 2 × 6s or as deep as 2 × 14s. Joists are spaced either 12 or 16 in. from adjacent joists.

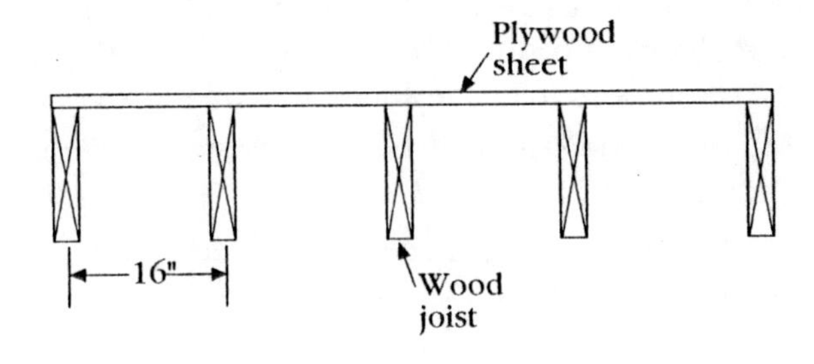

Figure 12-1–Elevation of a floor system

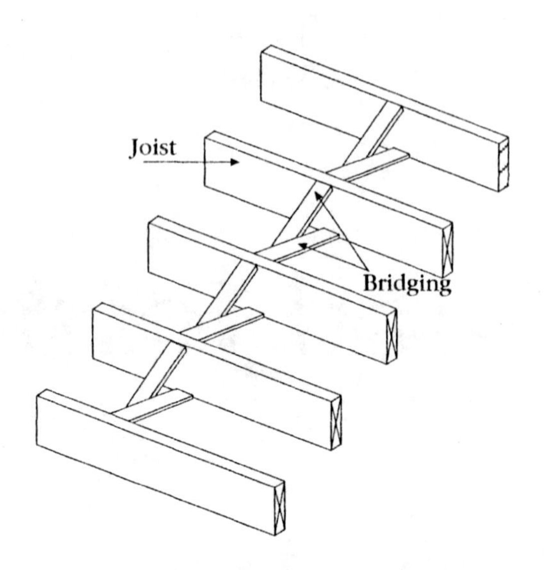

Figure 12-2–Isometric view of floor joists

Joists must be designed to support their own weight, the weight of the flooring material above as well as any load imposed by the occupants of the structure. If the space below the joists is inhabitable, the weight of a ceiling and other items that could be suspended from the underside of the joist must be considered in the design.

The loads that result from the weight of the permanent materials (joists, plywood flooring, floor covering, drywall, etc.) are classified as dead loads and are estimated to be approximately 10 psf to 15 psf. Less-permanent loads are classified as live loads. The required residential design live load is typically 30 psf for floors that support sleeping rooms. For other rooms, a 40-psf capacity is typically required. If heavy loads or concentrated loads are anticipated, the effects of these loads should be investigated (for example, a large whirlpool tub).

The design of a wood floor joist consists of checking that the joist satisfy the following criteria:

1. Deflection
2. Bending stress
3. Shear stress

Deflections

The deflection of the floor must be limited to avoid cracking of the drywall, unsightly sagging and most importantly, annoying bouncing of the floor when occupants are walking. The maximum permissible deflection for floor construction is:

$$\text{Maximum allowable deflection} = \text{span length}/360$$

Utilizing this equation, the maximum deflection allowed for a joist spanning 16 ft is:

$$\text{Maximum allowable deflection} = (16/360) = 0.044 \text{ ft} = 0.53 \text{ in.}$$

Therefore, the deflection caused by live loads on a 16-ft-long joist cannot exceed 1/2 in. Note that the only loads included in the deflection calculation are live loads. Only live-load deflection is of interest since they produce the forces that cause cracking, sagging and bouncing.

By the use of this allowable deflection calculation the allowable deflection for various spans can be determined. Results for these span lengths are shown in the following table:

Table 12-1—Maximum Allowable Deflections of Joists

Span Length (ft)	Maximum Allowable Deflection (in.)
10	0.33
12	0.40
14	0.47
16	0.53
18	0.60
20	0.67
22	0.73
24	0.80
26	0.87
28	0.93

Bending Stress

The bending stress is:

Bending stress = f = Md/2I

where M = Bending moment
d = Depth of the joist
I = Moment of inertia

The maximum bending moment allowed by code depends on the species, grade and size of the joist being used. The four following tables give the allowable bending stresses for four different species of wood. Three grades are given for each of the species. Select structural is a higher quality than type No. 1 and No. 2. No. 2 is the lowest quality of the three listed, with No. 1 being of intermediate quality.

Table 12-2—Allowable Joist-Bending Stress (psi)
Spruce–Pine–Fir

	Joist Depth					
	6	8	10	12	14	16
Select Structural	1869	1725	1581	1438	1294	1294
No. 1	1308	1208	1107	1006	906	906
No. 2	1308	1208	1107	1006	906	906

Table 12-3—Allowable Joist-Bending Stress (psi)
Hem–Fir

	Joist Depth					
	6	8	10	12	14	16
Select Structural	2093	1932	1771	1610	1449	1449
No. 1	1420	1311	1202	1093	983	983
No. 2	1271	1173	1075	978	880	880

**Table 12-4—Allowable Joist-Bending Stress (psi)
Northern Species**

	Joist Depth					
	6	8	10	12	14	16
Select Structural	1420	1311	1202	1093	983	983
No. 1	860	794	728	661	595	595
No. 2	860	794	728	661	595	595

**Table 12-5—Allowable Joist-Bending Stress (psi)
Southern Pine**

	Joist Depth					
	6	8	10	12	14	16
Select Structural	2128	2013	1725	1610	1450	1450
No. 1	1495	1380	1208	1121	1009	1009
No. 2	1323	1208	1064	1006	906	906

Shear Stress

The shear stress for a rectangular wood member is:

Shear stress = 3V/2bd

where: V = Shear
b = Width of joist
d = Depth of joist

The maximum allowable shear stress is a function of the species of wood and the amount of splits in the wood. The maximum allowable shear stresses for the previously discussed species of wood are given below. If the wood has minimal splits, these values could be increased by as much as a factor of 2.

Table 12-6—Allowable Shear Stresses (psi)

Species	Allowable Shear Stress (psi)
Spruce-Pine-Fir	70
Hem-Fir	75
Northern Species	65
Southern Pine	90

Example 1: Determine what species of wood would be acceptable for the following criteria for a joist.

- Depth of member 12 in.
- Shear stress resulting from loading = 50 psi
- Bending stress resulting from loading = 900 psi
- Joist = Grade No. 2
- Deflection resulting from loading = 0.5 in.
- Span length = 20 ft

<u>Check Spruce-Pine-Fir Species</u>

Allowable shear stress = 70 psi (which is greater than 50 psi)

Allowable bending stress = 1006 psi (which is greater than 900 psi)

Deflection = 0.5 (which is less than 0.67 in.)

∴ Spruce-Pine-Fir is acceptable

Check Hem-Fir Species

Allowable shear stress = 75 psi (which is greater than 50 psi)

Allowable bending stress = 973 psi (which is greater than 900 psi)

∴ Hem-Fir Group No. 2 is acceptable.

Check Northern Species

Allowable shear stress = 65 psi (which is greater than 50 psi)

Allowable bending stress = 667 psi (which is less than 900 psi)

∴ Northern Species Group No. 2 is unacceptable.

Check Southern Pine

Allowable shear stress = 90 psi (which is greater than 50 psi)

Allowable bending stress = 1006 psi (which is greater than 900 psi)

∴ Southern Pine Group No. 2 is acceptable.

Tables 12-7 through 12-10 (at the end of this topic) provide a tabular format for selecting joists when the design criteria is a uniform load. If the load, joist spacing and span length are known, the size joist required can be chosen from these tables. Defection, bending stress and shear stress criteria were all checked and satisfied in the construction of these tables.

Table 12-7—Allowable Span Lengths (Spruce–Pine–Fir)

Case 1: **Joist Spacing = 16 in.**
Live Load = 40 psf
Dead Load = 10 psf

	Span Length									
	10	12	14	16	18	20	22	24	26	28
Select Structural	2 × 8	2 × 8	2 × 10	2 × 10	2 × 12	2 × 14	2 × 14	2 × 16	2 × 16	–
No. 1	2 × 8	2 × 8	2 × 10	2 × 12	2 × 14	2 × 16	–	–	–	–
No. 2	2 × 8	2 × 8	2 × 10	2 × 12	2 × 14	2 × 16	–	–	–	–

Case 2: **Joist Spacing = 16 in.**
Live Load = 30 psf
Dead Load = 10 psf

	Span Length									
	10	12	14	16	18	20	22	24	26	28
Select Structural	2 × 6	2 × 8	2 × 10	2 × 10	2 × 12	2 × 12	2 × 14	2 × 14	2 × 16	2 × 16
No. 1	2 × 6	2 × 8	2 × 10	2 × 10	2 × 12	2 × 14	2 × 16	2 × 16	–	–
No. 2	2 × 6	2 × 8	2 × 10	2 × 10	2 × 12	2 × 14	2 × 16	2 × 16	–	–

Table 12-7—Allowable Span Lengths (Spruce–Pine–Fir) (cont.)

Case 3: **Joist Spacing = 12 in.**
Live Load = 40 psf
Dead Load = 10 psf

	Span Length									
	10	12	14	16	18	20	22	24	26	28
Select Structural	2 × 6	2 × 8	2 × 10	2 × 10	2 × 12	2 × 12	2 × 14	2 × 14	2 × 16	2 × 16
No. 1	2 × 6	2 × 8	2 × 10	2 × 10	2 × 12	2 × 14	2 × 16	2 × 16	–	–
No. 2	2 × 6	2 × 8	2 × 10	2 × 10	2 × 12	2 × 14	2 × 16	2 × 16	–	–

Case 4: **Joist Spacing = 12 in.**
Live Load = 30 psf
Dead Load = 10 psf

	Span Length									
	10	12	14	16	18	20	22	24	26	28
Select Structural	2 × 6	2 × 8	2 × 8	2 × 10	2 × 10	2 × 12	2 × 12	2 × 14	2 × 14	2 × 16
No. 1	2 × 6	2 × 8	2 × 8	2 × 10	2 × 10	2 × 12	2 × 14	2 × 14	2 × 16	2 × 16
No. 2	2 × 6	2 × 8	2 × 8	2 × 10	2 × 10	2 × 12	2 × 14	2 × 14	2 × 16	2 × 16

Table 12-8—Allowable Span Lengths (Hem–Fir)

Case 1: **Joist Spacing = 16 in.**
Live Load = 40 psf
Dead Load = 10 psf

	Span Length									
	10	12	14	16	18	20	22	24	26	28
Select Structural	2 × 8	2 × 8	2 × 10	2 × 10	2 × 12	2 × 14	2 × 14	2 × 16	2 × 16	–
No. 1	2 × 8	2 × 8	2 × 10	2 × 12	2 × 14	2 × 16	2 × 16	–	–	–
No. 2	2 × 8	2 × 10	2 × 10	2 × 12	2 × 14	2 × 16	–	–	–	–

Case 2: **Joist Spacing = 16 in.**
Live Load = 30 psf
Dead Load = 10 psf

	Span Length									
	10	12	14	16	18	20	22	24	26	28
Select Structural	2 × 6	2 × 8	2 × 8	2 × 10	2 × 10	2 × 12	2 × 14	2 × 14	2 × 16	2 × 16
No. 1	2 × 6	2 × 8	2 × 10	2 × 10	2 × 12	2 × 14	2 × 14	2 × 16	–	–
No. 2	2 × 6	2 × 8	2 × 10	2 × 10	2 × 12	2 × 14	2 × 16	2 × 16	–	–

Table 12-8—Allowable Span Lengths (Hem–Fir) (cont.)

Case 3: **Joist Spacing = 12 in.**
Live Load = 40 psf
Dead Load = 10 psf

	Span Length									
	10	12	14	16	18	20	22	24	26	28
Select Structural	2 × 6	2 × 8	2 × 8	2 × 10	2 × 10	2 × 12	2 × 14	2 × 14	2 × 16	2 × 16
No. 1	2 × 6	2 × 8	2 × 10	2 × 10	2 × 12	2 × 12	2 × 14	2 × 16	2 × 16	–
No. 2	2 × 6	2 × 8	2 × 10	2 × 10	2 × 12	2 × 14	2 × 16	2 × 16	–	–

Case 4: **Joist Spacing = 12 in.**
Live Load = 30 psf
Dead Load = 10 psf

	Span Length									
	10	12	14	16	18	20	22	24	26	28
Select Structural	2 × 6	2 × 8	2 × 8	2 × 10	2 × 10	2 × 12	2 × 12	2 × 12	2 × 14	2 × 14
No. 1	2 × 6	2 × 8	2 × 8	2 × 10	2 × 10	2 × 12	2 × 12	2 × 14	2 × 16	2 × 16
No. 2	2 × 6	2 × 8	2 × 8	2 × 10	2 × 10	2 × 12	2 × 14	2 × 14	2 × 16	–

Table 12-9—Allowable Span Lengths (Northern Species)

Case 1: **Joist Spacing = 16 in.**
Live Load = 40 psf
Dead Load = 10 psf

	Span Length									
	10	12	14	16	18	20	22	24	26	28
Select Structural	2 × 8	2 × 10	2 × 10	2 × 12	2 × 14	2 × 16	2 × 16	–	–	–
No. 1	2 × 10	2 × 12	2 × 14	2 × 16	–	–	–	–	–	–
No. 2	2 × 10	2 × 12	2 × 14	2 × 16	–	–	–	–	–	–

Case 2: **Joist Spacing = 16 in.**
Live Load = 30 psf
Dead Load = 10 psf

	Span Length									
	10	12	14	16	18	20	22	24	26	28
Select Structural	2 × 8	2 × 8	2 × 10	2 × 12	2 × 12	2 × 14	2 × 14	2 × 16	–	–
No. 1	2 × 8	2 × 10	2 × 12	2 × 14	2 × 16	–	–	–	–	–
No. 2	2 × 8	2 × 10	2 × 12	2 × 14	2 × 16	–	–	–	–	–

Table 12-9—Allowable Span Lengths (Northern Species) (cont.)

Case 3: **Joist Spacing = 12 in.**
Live Load = 40 psf
Dead Load = 10 psf

	Span Length									
	10	12	14	16	18	20	22	24	26	28
Select Structural	2 × 8	2 × 8	2 × 10	2 × 12	2 × 12	2 × 14	2 × 14	2 × 16	2 × 16	–
No. 1	2 × 8	2 × 10	2 × 12	2 × 14	2 × 16	2 × 14	–	–	–	–
No. 2	2 × 8	2 × 10	2 × 12	2 × 14	2 × 16	2 × 14	–	–	–	–

Case 4: **Joist Spacing = 12 in.**
Live Load = 30 psf
Dead Load = 10 psf

	Span Length									
	10	12	14	16	18	20	22	24	26	28
Select Structural	2 × 6	2 × 8	2 × 10	2 × 10	2 × 12	2 × 12	2 × 14	2 × 14	2 × 16	2 × 16
No. 1	2 × 8	2 × 8	2 × 10	2 × 12	2 × 14	2 × 16	2 × 16	–	–	–
No. 2	2 × 8	2 × 8	2 × 10	2 × 12	2 × 14	2 × 16	2 × 16	–	–	–

Table 12-10—Allowable Span Lengths (Southern Pine)

Case 1: **Joist Spacing = 16 in.**
Live Load = 40 psf
Dead Load = 10 psf

	Span Length									
	10	12	14	16	18	20	22	24	26	28
Select Structural	2 × 8	2 × 8	2 × 10	2 × 10	2 × 12	2 × 14	2 × 14	2 × 16	2 × 16	–
No. 1	2 × 8	2 × 8	2 × 10	2 × 12	2 × 14	2 × 16	2 × 16	–	–	–
No. 2	2 × 8	2 × 10	2 × 10	2 × 12	2 × 14	2 × 16	–	–	–	–

Case 2: **Joist Spacing = 16 in.**
Live Load = 30 psf
Dead Load = 10 psf

	Span Length									
	10	12	14	16	18	20	22	24	26	28
Select Structural	2 × 6	2 × 8	2 × 8	2 × 10	2 × 10	2 × 12	2 × 14	2 × 14	2 × 16	2 × 16
No. 1	2 × 6	2 × 8	2 × 10	2 × 10	2 × 12	2 × 14	2 × 14	2 × 16	–	–
No. 2	2 × 6	2 × 8	2 × 10	2 × 10	2 × 12	2 × 14	2 × 16	2 × 16	–	–

Table 12-10—Allowable Span Lengths (Southern Pine) (cont.)

Case 3: **Joist Spacing = 12 in.**
Live Load = 40 psf
Dead Load = 10 psf

	Span Length									
	10	12	14	16	18	20	22	24	26	28
Select Structural	2 × 6	2 × 8	2 × 8	2 × 10	2 × 10	2 × 12	2 × 14	2 × 14	2 × 16	2 × 16
No. 1	2 × 6	2 × 8	2 × 10	2 × 10	2 × 12	2 × 12	2 × 14	2 × 16	2 × 16	–
No. 2	2 × 6	2 × 8	2 × 10	2 × 10	2 × 12	2 × 14	2 × 16	2 × 16	–	–

Case 4: **Joist Spacing = 12 in.**
Live Load = 30 psf
Dead Load = 10 psf

	Span Length									
	10	12	14	16	18	20	22	24	26	28
Select Structural	2 × 6	2 × 8	2 × 8	2 × 10	2 × 10	2 × 12	2 × 12	2 × 12	2 × 14	2 × 14
No. 1	2 × 6	2 × 8	2 × 8	2 × 10	2 × 10	2 × 12	2 × 12	2 × 14	2 × 16	2 × 16
No. 2	2 × 6	2 × 8	2 × 8	2 × 10	2 × 10	2 × 12	2 × 14	2 × 14	2 × 16	2 × 16

Advanced Discussion

In the introduction portion of the discussion of wood joists the proper size joist to adequately support the imposed loads was determined by selecting the joist size from the appropriate table. The values that comprise these tables were calculated by considering the limitations set by the building code for deflection, bending stress and shear. This Advanced Discussion provides some detail for the calculations necessary to determine the joist size, as opposed to selecting the joist size from the tables.

The deflection of the floor must not exceed the span length divided by 360 to be in conformance with the building code. Remember that the deflection criteria applies only to the effects of live loads since live loads are the cause of deflection-related problems. The maximum deflection for a simply supported beam under a uniformly distributed load is:

$$\text{Maximum deflection} = 5w\ell^4/EI$$

where
w = Uniformly distributed live load
ℓ = Span length
E = Modulus of elasticity
I = Moment of inertia

The modulus of elasticity for the species commonly used for joists is given in the following table for different species and grades of lumber.

Table 12-11—Modulus of Elasticity, E (psi)

	Select Structural	No. 1	No. 2
Spruce-Pine-Fir	1,500,000	1,400,000	1,400,000
Hem-Fir	1,600,000	1,500,000	1,300,000
Northern Species	1,100,000	1,100,000	1,100,000
Southern Pine	1,600,000	1,500,000	1,400,000

Example 3: Determine the midspan live load deflection of a 2 × 12 ($I = 178$ in.^{4}) floor joist that spans 16 ft (192 in.). Assume a live load of 40 psf, a dead load of 10 psf and a 16-in. joist spacing. The joist is southern pine ($E = 1{,}400{,}000$ psi), Group No. 2.

$$\text{Maximum deflection} = 5w\ell^4/EI$$

$$\text{Maximum deflection} = \frac{5 \times w \times (192)^4}{384 \times 1{,}400{,}000 \quad 178}$$

$$= 0.071w$$

The live load is given as 40 lb/ft^2 (psf) which is equivalent to 0.277 lb/in.^{2} (psi) (40 psf/144 in.^{2} = 0.277 psi). Since the joists are spaced at 16 in., each joist will be required to carry a load of a 16-in. width of floor. The applied uniform load to each joist is actually 16 × 0.277 = 4.44 lb/in.. Therefore the maximum deflection is:

$$\begin{aligned}\text{Maximum deflection} &= 0.071w \\ &= 0.071 \times 4.44 \\ &= 0.315 \text{ in.}\end{aligned}$$

The maximum allowable deflection for this floor joist is:

$$\begin{aligned}\text{Maximum allowable deflection} &= \ell/360 \\ &= 192/360 \\ &= 0.533 \text{ in.}\end{aligned}$$

The maximum allowable deflection is 0.533 in. Therefore the beam is satisfactory with respect to deflection requirements. Note that if the deflection was calculated for a 2 × 10 under the same loading conditions it would be equal to 0.56 in. which would have been in excess of the allowable and unacceptable.

The bending stress caused by the loads is determined using the following equation:

$$\text{Bending stress} = Md/2I$$

where M = Bending moment
d = Depth of joist
I = Moment of inertia

The maximum bending moment, M, for a joist with a uniform load is determined by the following equation:

$$M = w\ell^2/8$$

Note that the uniform load, w, that is inserted into this equation consists of both the dead and live loads since they both produce bending stresses. Therefore w equals 50 psf (10 psf + 40 psf = 50 psf), which is equivalent to 0.347 psi (50 psf/144 in.^{2} = 0.347 psi). The applied uniform load to each joist is actually 16 × 0.347 = 5.56 lb/in. The bending moment is then calculated as:

$$M = w\ell^2/8$$

$$\begin{aligned} M &= 5.55 \times (192)^2/8 \\ &= 25{,}600 \text{ in.}^2/\text{in.-lb} \end{aligned}$$

Therefore, the bending stress is:

$$\text{Bending stress} = Md/2I$$

$$\text{Bending stress} = \frac{25{,}600 \times 11.25}{2 \times 178}$$

$$= 809 \text{ psi}$$

Table 12-5 from the Introduction portion of this topic provided the allowable bending stress for various sizes and grades of southern pine lumber. This is the table for southern pine fir.

	Joist Depth					
	6	8	10	12	14	16
Select Structural	2128	2013	1725	1610	1450	1450
No. 1	1495	1380	1208	1121	1009	1009
No. 2	1323	1208	1064	1006	906	906

Review of the table reveals that the allowable bending stress for a 2 × 12, Grade No. 2 is 1006 psi. This allowable bending stress exceeds the actual bending stress of 809 psi that is produced by the loads.

The shear stress for a rectangular wood section is calculated:

Shear stress = 3V/bd

where V = Shear
b = Width of joist
d = Depth of joist

Joists which are uniformly loaded are rarely controlled by shear stresses. The controlling factor is usually deflection constraints, however it is not unusual that bending stresses may control.

The shear, V, for a uniformly loaded joist is:

$$V = w\ell/2$$

Using the same values as in the previous example, the shear is:

$$V = w\ell/2$$

$$V = 5.56 \times 192/2$$
$$= 534 \text{ lb}$$

The shear stress is:

Shear stress = 3V/bd

$$\text{Shear stress} = \frac{3 \times 534}{2 \times 1.5 \quad 11.25}$$

$$= 47.5 \text{ psi}$$

This is less than the allowable shear stress for southern pine of 90 psi.

Therefore a 2 × 12 is adequate for the imposed loading condition. Note that a 2 × 10 would have satisfied the load criteria in regard to bending stress and shear stress. However, it would not have satisfied deflection criteria. This implies that the beam would probably have significant bounce during service but would not fail.

13

Wood Trusses

Introduction

A truss is one of the most economical load-carrying members. The triangular shapes that compose the truss provide a cost-effective way to span long distances. An even greater savings is realized when many identical trusses are used. This repetitive construction results in reduced fabrication and installation costs.

Trusses are limited to use as the roof structure in residential construction. These members are constructed of wood, spaced on 2 to 3-ft centers, and span from 15 to 40 ft.

In its purest form a truss is composed of short members which are connected together at junctions called *panel points*. The connections at these joints are considered to be hinged. That is, the joint will transfer axial forces only and will be unable to resist any rotational forces. An everyday example of a hinge in a truss is a door hinge. This hinge supports the weight of the door but yet does not resist the circular motion to open it. Therefore, a hinge in a truss allows for the transfer of forces but does not resist end rotations.

This hinged connection concept allows for an uncomplicated analysis of the truss. A considerable amount of time in introductory structural analysis courses is spent on analysis of trusses with hinged connections. This type of analysis is helpful in explaining the basic concept of structural analysis. Anyone who has examined a truss knows that the connection at panel points are not hinged and do resist some rotation. This reduces the validity of the hinged connection truss analysis and requires a more complicated analysis. In fact, a thorough design of a wood truss is usually facilitated by a very complicated and detailed computer software. This software, in most cases is developed by companies that fabricate trusses. This allows for the truss supplier to provide complete design and fabrication service. Figure 13-1 provides the names of the various components of a truss.

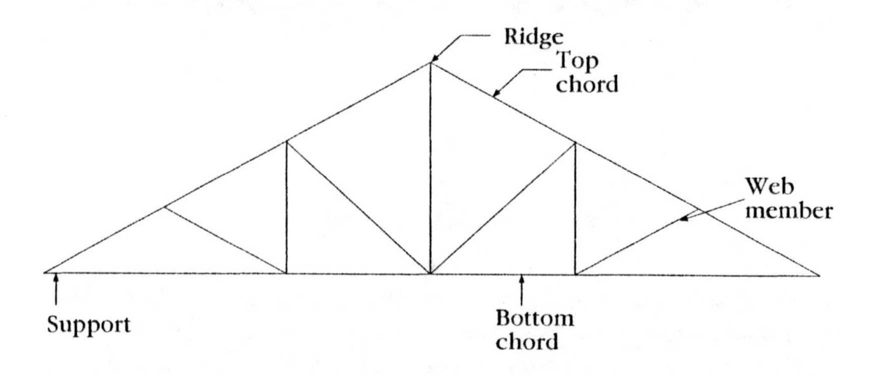

Figure 13-1—Components of a truss

Trusses fall into three categories, as shown in Figure 13-2. These are pitched, flat and bowstring. Typically only pitched trusses are used in residential construction. Pitched trusses can be further divided into 30 or more types. The four most popular types are shown in Figure 13-3. The span ranges that each of these trusses are used for are:

Queen truss	15 to 25 ft
Fink truss	25 to 35 ft
Fan truss	35 to 40 ft
Double fink truss	40 ft and larger

Two other trusses that are often used under special conditions are shown in Figure 13-4. The scissors truss is used when a sloped ceiling on the interior is desired. The piggyback truss is utilized when the overall height of the truss is in excess of 11 or more feet. This truss height causes delivery problems since the truss overhangs the sides of the delivery truck. Other trusses that are available but not commonly used are shown in Figure 13-5.

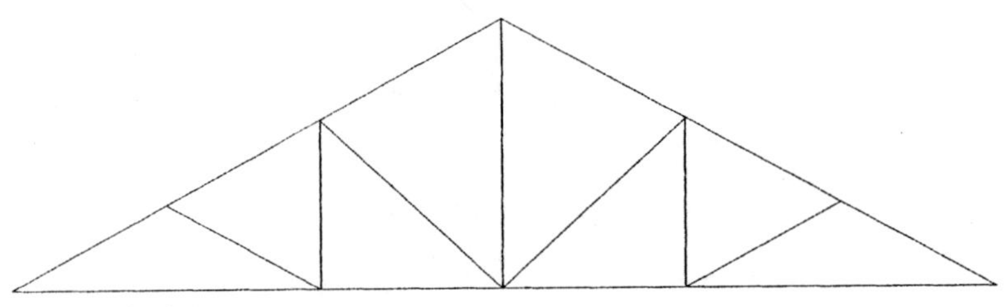

2a. Pitched truss

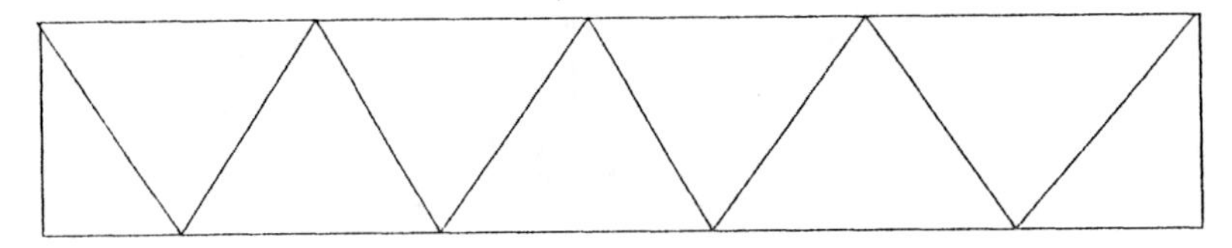

2b. Flat truss

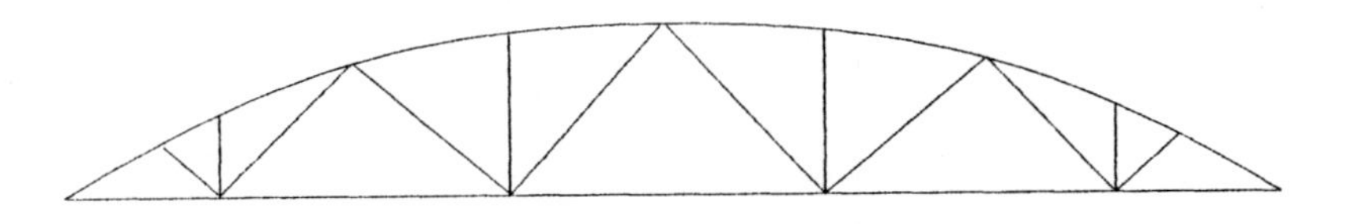

2c. Bowstring truss

Figure 13-2—Truss designation by shape

Trusses must be designed to withstand a variety of loads. These loads include the following:

- Weight of the truss
- Weight of the shingles and sheathing
- Weight of drywall attached to bottom chord
- Snow loads on roof
- Live loads on roof
- Wind loads on roof
- Loads from attic storage

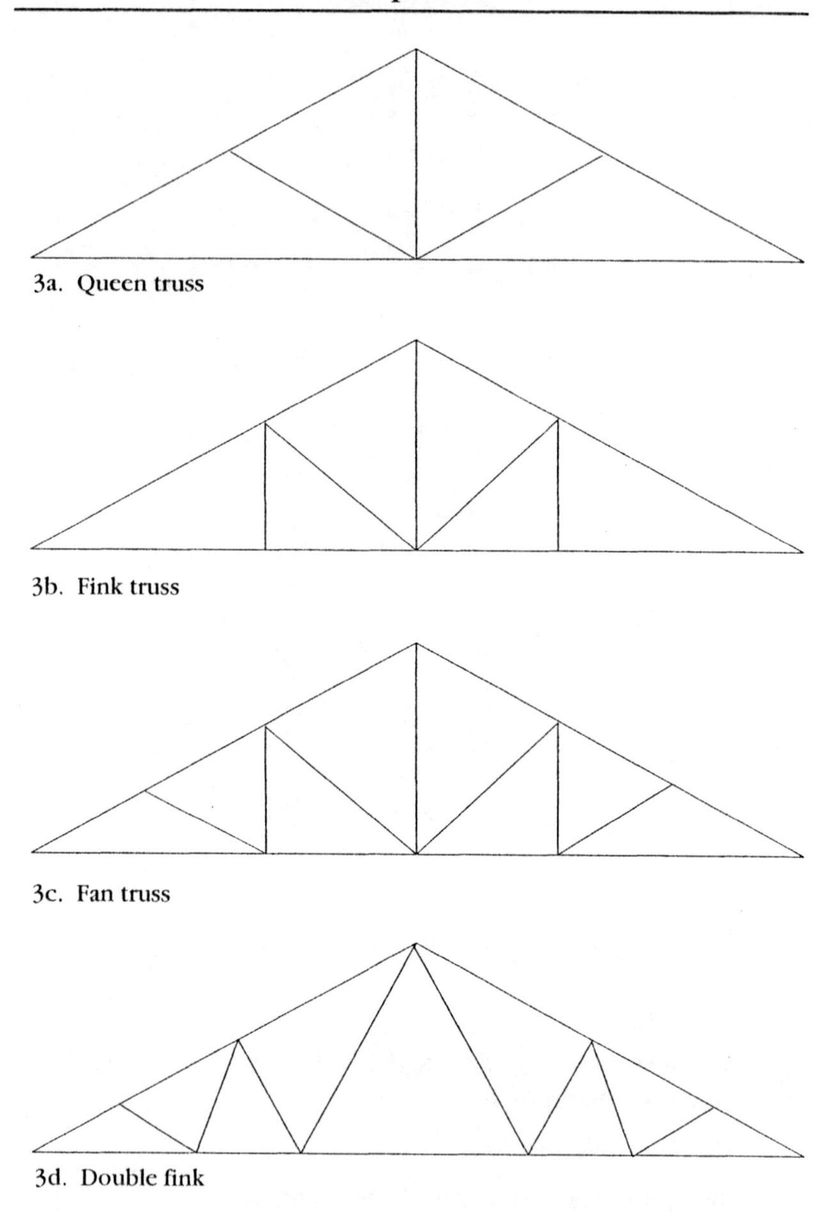

Figure 13-3—Trusses typically used in residential construction

The list of loads are for in-service loading conditions. Equally, if not more important, are loads produced during construction. These include transporting the prefabricated truss, lifting it into place and stabilizing it prior to sheathing insulation. Damage to

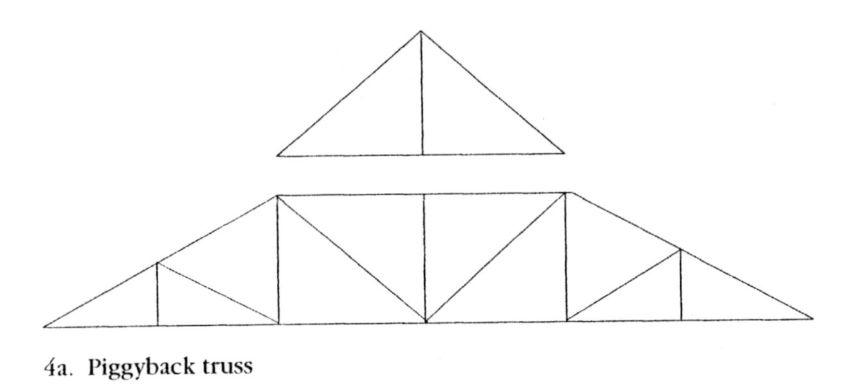

4a. Piggyback truss

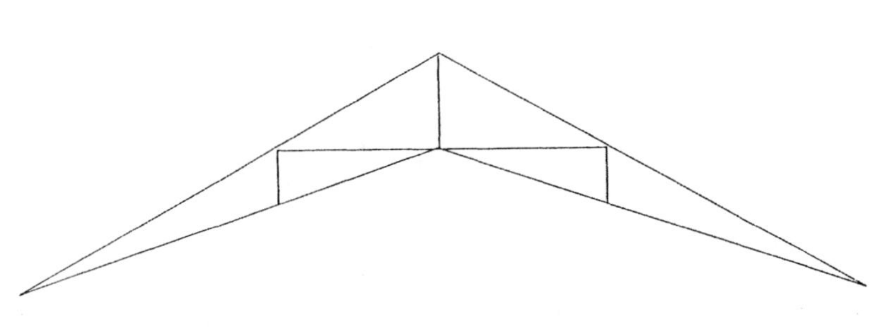

4b. Scissors truss

Figure 13-4—Trusses for long-span and cathedral ceilings

the truss, or allowing it to remain unstable, can lead to disastrous results.

When prefabricated trusses are loaded on the truck for shipment, care must be taken to avoid deforming any portion of the truss that overhangs the truck trailer. Upon arrival to the jobsite, it is optimal to have the trusses unloaded from the truck directly on to the roof. This avoids having to store the trusses and also having to move them more than once. If the trusses cannot be immediately installed, they should be stored in such a manner that they will not be damaged. During installation of the trusses, bracing must be provided. The Truss Plate Institute located in Madison, Wisconsin has several easy-to-understand publications on this topic.

Trusses are braced to prevent them from tipping over. Bracing a truss to an adjacent truss usually is not sufficient because the truss

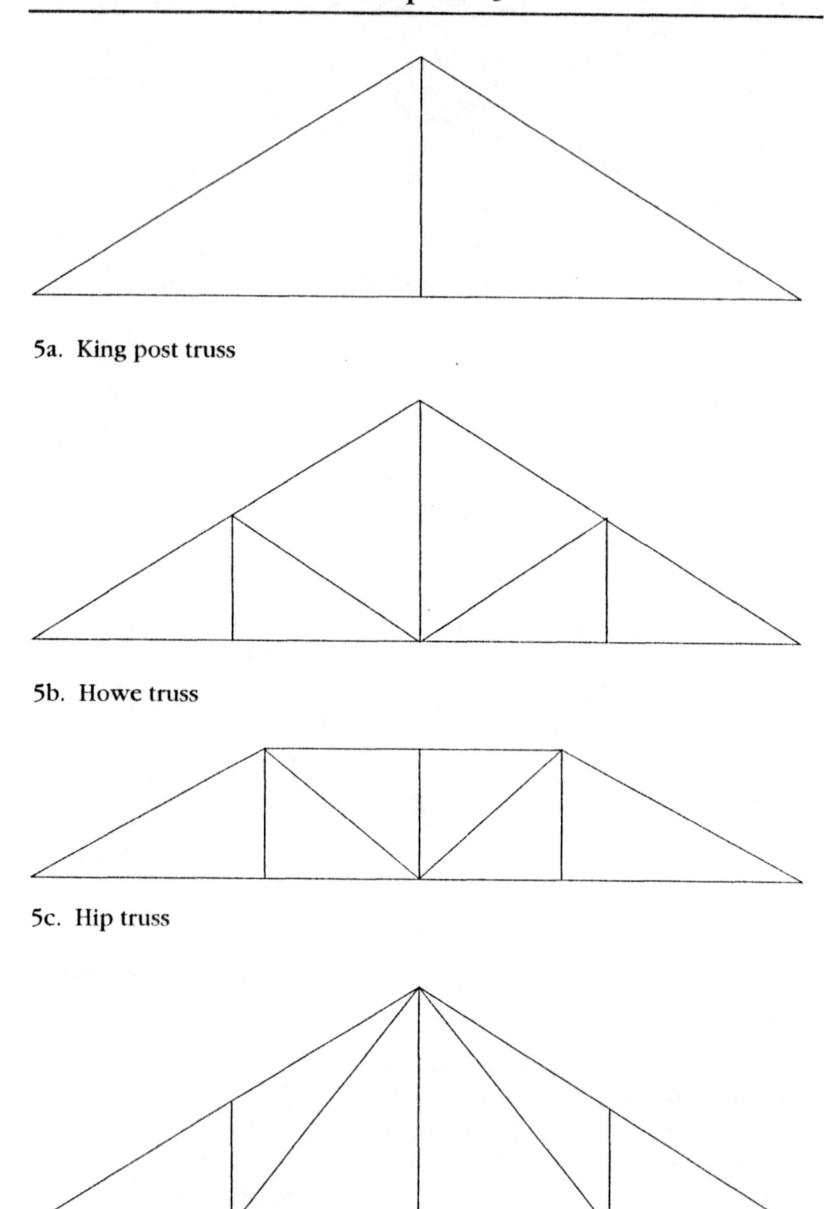

Figure 13-5—Trusses available in construction industry

might topple over together. It is usually recommended that at least one truss be braced to a solid element such as a wall or to the ground.

Care must also be taken to avoid overloading the truss with the roofing material. Roof sheathing should not be stored all at one location on the trusses. This heavy concentrated load can invite collapse or permanently deform the truss.

Advanced Discussion

As previously noted, the analysis of a truss is complicated and is often performed with the assistance of a computer. Simplified analyses performed without a computer, although not totally correct, can still produce results that are fairly close to the more exact results. The results presented in the following discussion are based on simplified analysis.

Four tables are presented in the following discussion. These tables present the resulting forces in the members shown in the drawing accompanying the table. Results are given for a queen truss, fink truss, fan truss and double fink truss. The length of each truss is:

Queen truss	20 ft
Fink truss	24 ft
Fan truss	30 ft
Double fink truss	40 ft

The results in Table 13-1 provides the forces in the members for a roof angle from 10 degrees to 45 degrees. The designation, (T), signifies that the member is in tension. The designation, (C), signifies that the member is in comparison.

The table shows that the forces are increased by a factor of 4 when the angle of the truss is reduced from 45 degrees to 10 degrees. As expected, in this table and all the other tables, the top chord is in compression and the bottom chord is in tension.

The results presented in Table 13-2 are for the fink truss. The angles of the truss presented range from 10 degrees to 40 degrees. The angle is limited to 40 degrees, since beyond this angle the truss height at the peak is excessive. With the exception of the web

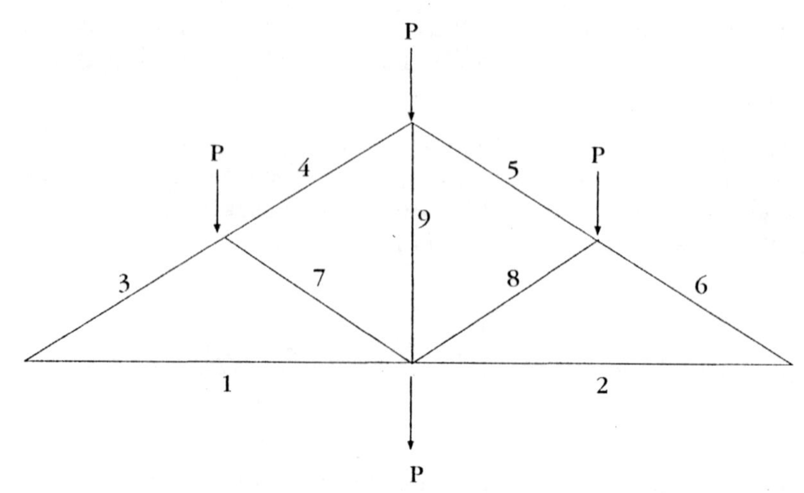

Table 13-1 Queen Truss–Multiplier Times Load P for 20-ft Span

Member Number	Roof Angle				
	10°	20°	30°	40 °	45°
1 (2 also)	11.3(T)	5.5(T)	3.5(T)	2.4(T)	2.0(T)
3 (6 also)	11.5(C)	5.8(C)	4.0(C)	3.1(C)	2.8(C)
4 (5 also)	8.6(C)	4.4(C)	3.0(C)	2.3(C)	2.1(C)
7 (8 also)	2.9(C)	1.5(C)	1.0(C)	6.8(C)	0.7(C)
9	2.0(T)	2.0(T)	2.0(T)	2.0(T)	2.0(T)
End Reaction	2.0	2.0	2.0	2.0	2.0

members, the forces in the members increase by a factor of 4 when the angle is decreased from 40 degrees to 10 degrees.

The results presented in Table 13-3 for the fan truss are for angles from 10 degrees to 35 degrees. Results are similar to the previous two trusses discussed. The results for a double fink truss are shown in Table 13-4. With the exception of web members, the force in the members doubles when the angle is decreased from 25 to 10 degrees.

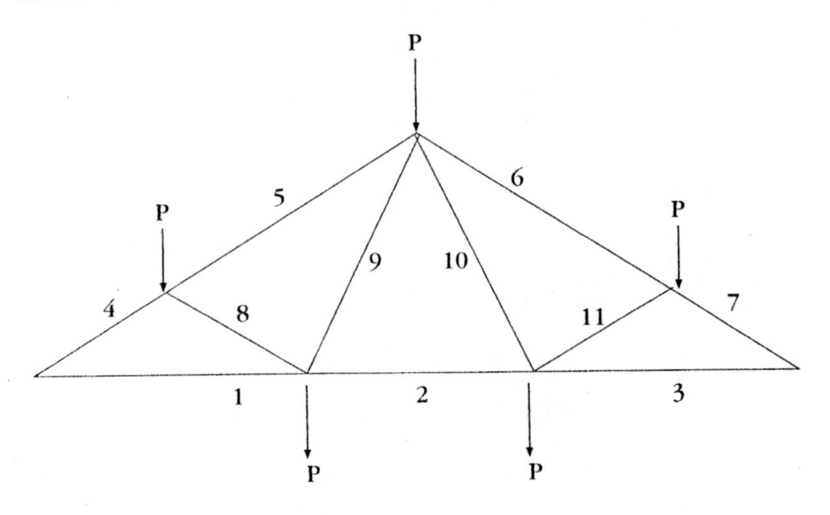

Table 13-2 Fink Truss–Multiplier Times Load P for 24-ft Span

Member Number	Roof Angle			
	10°	20°	30°	40°
1 (3 also)	14.2(T)	6.9(T)	4.3(T)	3.0(T)
2	8.5(T)	4.1(T)	2.6(T)	1.8(T)
4 (7 also)	14.4(C)	7.3(C)	5.0(C)	3.9(C)
5 (6 also)	11.5(C)	5.8(C)	4.0(C)	3.1(C)
8 (11 also)	2.9(C)	1.5(C)	1.0(C)	0.8(C)
9 (10 also)	3.2(T)	2.0(T)	1.7(T)	1.6(T)
End Reaction	2.5	2.5	2.5	2.5

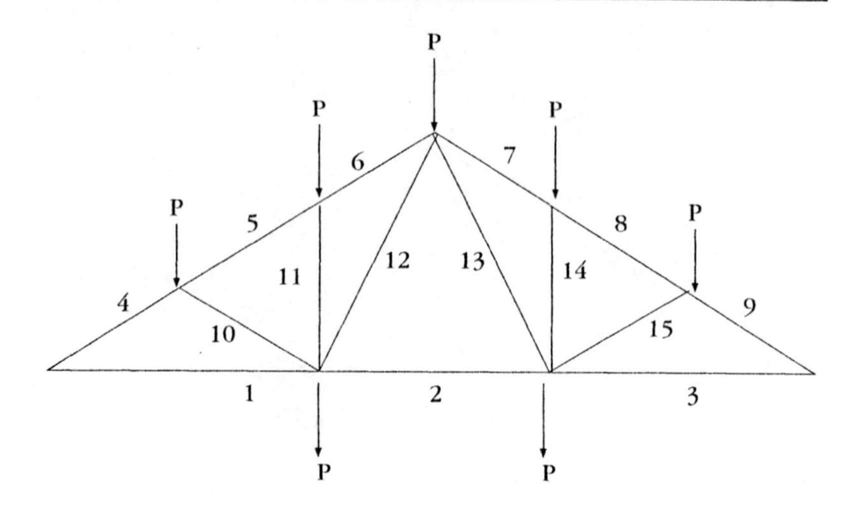

Table 13-3 Fan Truss–Multiplier Times Load P for 30-ft Span

Member Number	Roof Angle				
	10°	20°	30°	40°	45°
1 (2 also)	11.3(T)	5.5(T)	3.5(T)	2.4(T)	2.0(T)
3 (6 also)	11.5(C)	5.8(C)	4.0(C)	3.1(C)	2.8(C)
4 (5 also)	8.6(C)	4.4(C)	3.0(C)	2.3(C)	2.1(C)
7 (8 also)	2.9(C)	1.5(C)	1.0(C)	6.8(C)	0.7(C)
9	2.0(T)	2.0(T)	2.0(T)	2.0(T)	2.0(T)
End Reaction	2.0	2.0	2.0	2.0	2.0

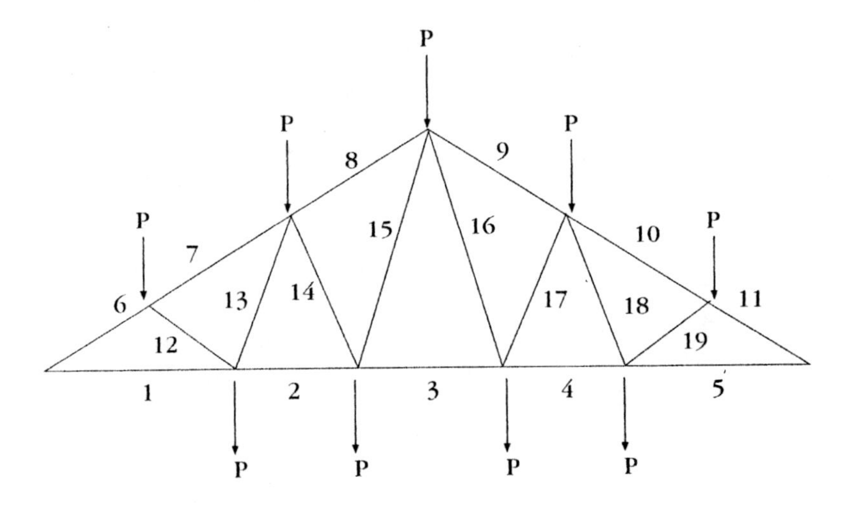

Table 13-4 Double Fink–Multiplier Times Load P for 40-ft Span

Member Number	Roof Angle			
	10°	15°	20°	25°
1 (5 also)	25.5(T)	16.8(T)	12.4(T)	9.7(T)
2 (4 also)	22.7(T)	14.9(T)	11.0(T)	8.6(T)
3	14.2(T)	9.3(T)	6.9(T)	5.4(T)
6 (11 also)	25.9(C)	17.4(C)	13.2(C)	10.6(C)
7 (10 also)	23.0(C)	15.5(C)	11.7(C)	9.5(C)
8 (9 also)	20.2(C)	13.0(C)	9.6(C)	7.6(C)
12 (19 also)	2.9(C)	1.9(C)	1.5(C)	1.2(C)
13 (18 also)	3.6(T)	2.4(T)	2.0(T)	1.8(T)
14 (17 also)	2.0(C)	1.9(C)	1.9(C)	1.8(C)
15 (16 also)	6.3(T)	4.2(T)	3.3(T)	2.9(T)
End Reaction	4.5	4.5	4.5	4.5

14

Wood Headers

Introduction

Roof loads (snow, ice, dead loads, live loads, etc.) and the loads applied to the floors (e.g. live loads and dead loads) are carried by the roof and floor joists to the walls. The walls, which are constructed of wood studs, carry the loads to the foundation. This load-path scenario is rather simple. This simplicity is lost when the wall framing is interrupted by openings in the framing for windows and doors. In these cases, the opening must be constructed so that the loads are transferred down to the foundation.

The carrying of the loads over the opening of a window or door is accomplished by using a header. A header consists of one or more beams of a substantial size to carry the loads applied at the opening. The header must be strong enough to carry the loads and stiff enough to avoid unacceptable deflections.

Various examples of locations where headers are used are shown in Figures 14-1 through 14-3. Figure 14-1 depicts a header used to span over the opening for a garage door. Spans for garage door openings range from 7 to 16 ft in length. Figure 14-2 depicts a header used over the opening for a window. In this case, the header carries loads from the roof, wall and floor.

Figure 14-3 also depicts a header used over a window opening. In this case, the header carries only wall loads. The roof and floor loads are carried to the front and back wall.

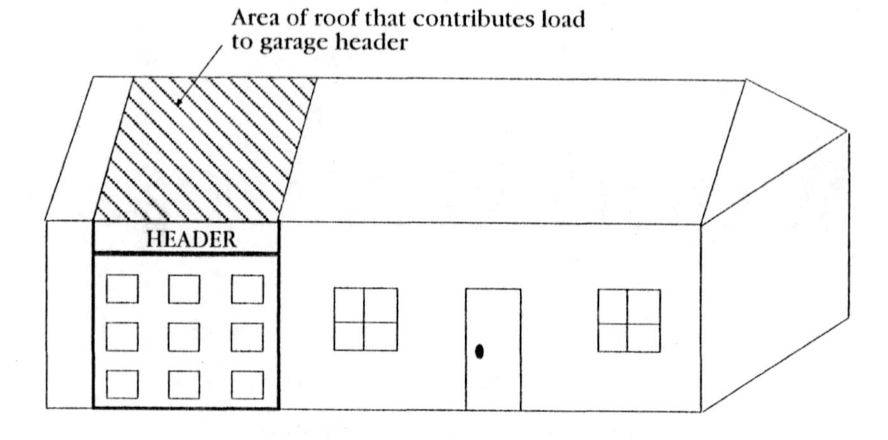

1a. Garage overhead door header–isometric view

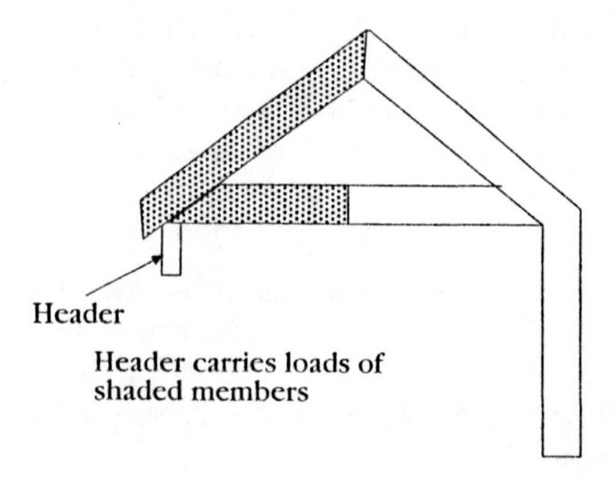

1b. Garage overhead door header–section view

Figure 14-1—Garage header

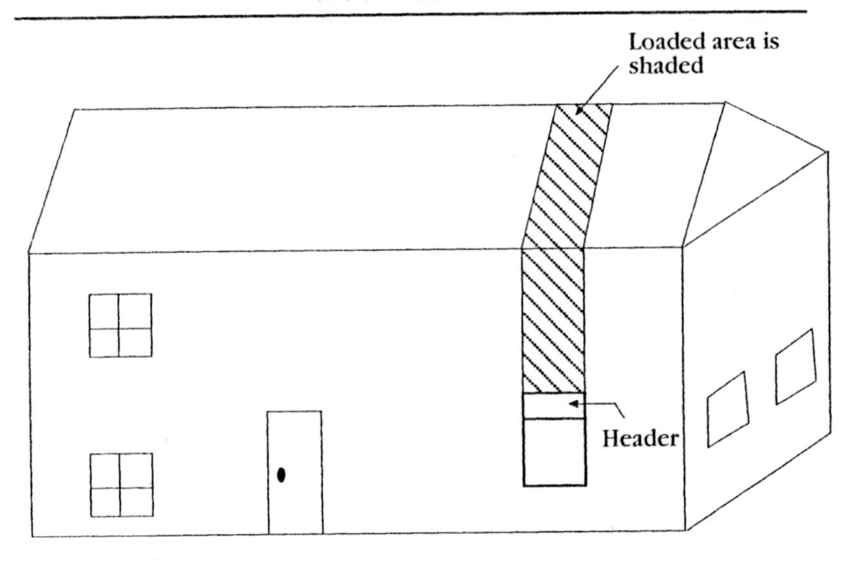

2a. Load on window header–isometric view

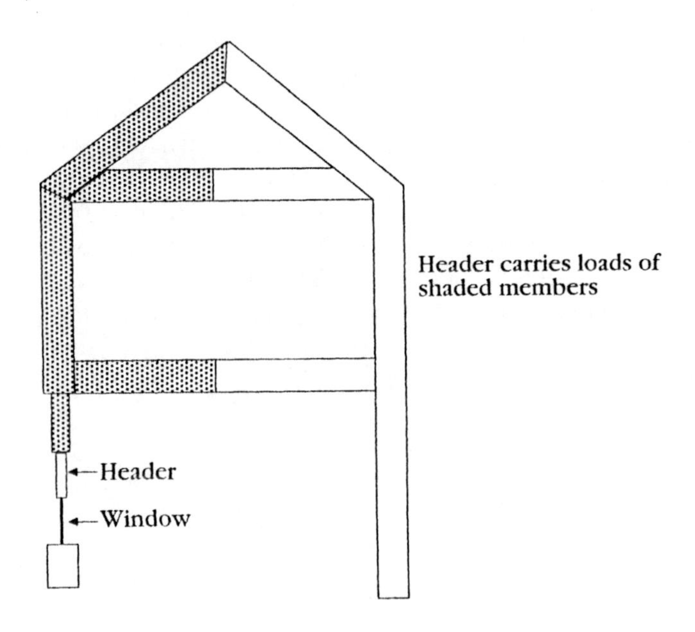

2b. Load on window header–section view

Figure 14-2—Window header

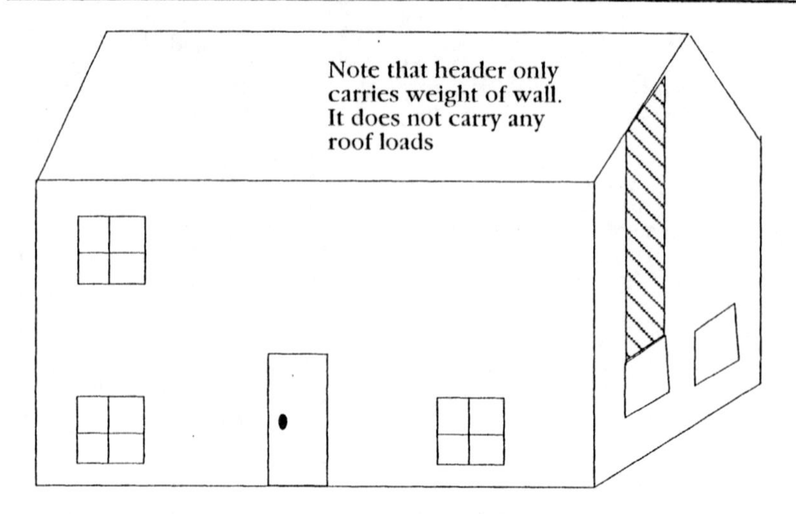

Figure 14-3—Load on window on gable side of house–isometric

A header must be properly sized to carry the applied loads and fit practically into the construction of the building. An opening in the building for a window or door will usually have 2 × 4 framing members. In particular, the studs on the sides of the window will be 2 × 4s that will give a wall thickness of 3-1/2 in. Headers are constructed of 2 × 6, 2 × 8, 2 × 10, 2 × 12, and 2 × 14 members. These members are 1-1/2 in. thick. Assuming two headers are side by side in the opening, the headers would be 3 in. wide, as compared to a 3-1/2-in. wall thickness. This 1/2-in. difference is compensated by adding a 1/2-in.-thick piece of plywood between the headers to obtain a member with an overall thickness equal to the wall thickness. The detail showing the double header is shown in Figure 14-4.

Headers are designed for the following items:

1. Deflection
2. Bending
3. Shear
4. Compression at support

As presented in Chapter 12 ("Wood Joists"), the allowable deflection depends on the joists used and whether it is a temporary or a permanent load condition. The allowable deflections for a header supporting a roof is between 33% and 50% greater than the header

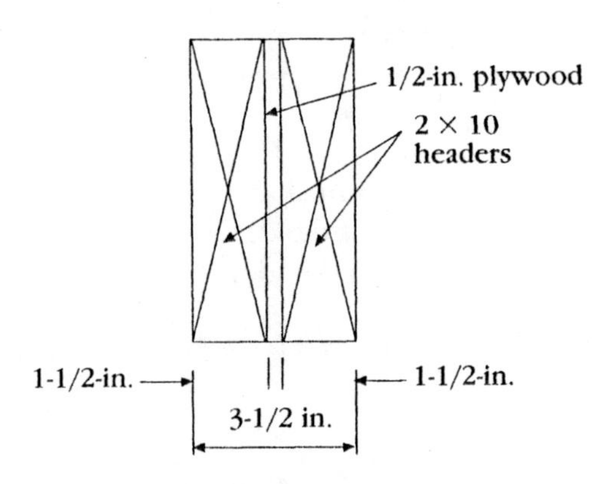

Figure 14-4—Cross-section of header of two 2 × 10 headers with a 1/2-in. plywood flitch plate

supporting floor loads. Likewise, allowable deflection requirements are less stringent for short-term loads as opposed to long term loads. Typical requirements for floor headers are:

$$\text{Total load} = \ell/240$$

$$\text{Live load} = \ell/360$$

For roofs, the allowable deflection is given by:

$$\text{Total load} = \ell/180$$

$$\text{Live load} = \ell/240$$

Bending moment stresses are calculated as previously discussed:

$$f = Md/2I$$

Shear stresses are much more predominate for headers as opposed to joists. This is because they have relatively high loads over short spans. It is important to note that loads within a distance from supports equal to the depth of the member can be ignored. This can be accomplished by decreasing the shear by multiplying by the following ratio for uniformly loaded spans.

$$\text{Multiplier} = 1 - 2d/\ell$$

The shear stress is calculated as:

$$v = 1.5V/bd$$

where V = Maximum shear
b = Width of header
d = Depth of header

The final item that must be checked is compression stresses at the supports. This is determined as:

$$\text{Compression stresses} = \frac{\text{Support force}}{\text{Area of support}}$$

The area of support is equal to the area of the top of the studs that the header is supported on. The area of a 2×4 stud is:

$$\text{Area} = 1.5 \times 3$$
$$= 4.5 \text{ in.}$$

Example 1

Determine the largest opening that two 2×12 headers can span for window "A" in the figure.

Assume the following:

Roof snow load = 20 psf

Floor live load = 40 psf

Attic live load = 30 psf

E = 1,600,000 psi

I = Two 2×12's

$= 178 \times 2 = 356 \text{ in.}^4$

Allowable bending stress = 1000 psi

Allowable shear stress = 90 psi

Allowable bearing stress = 600 psi

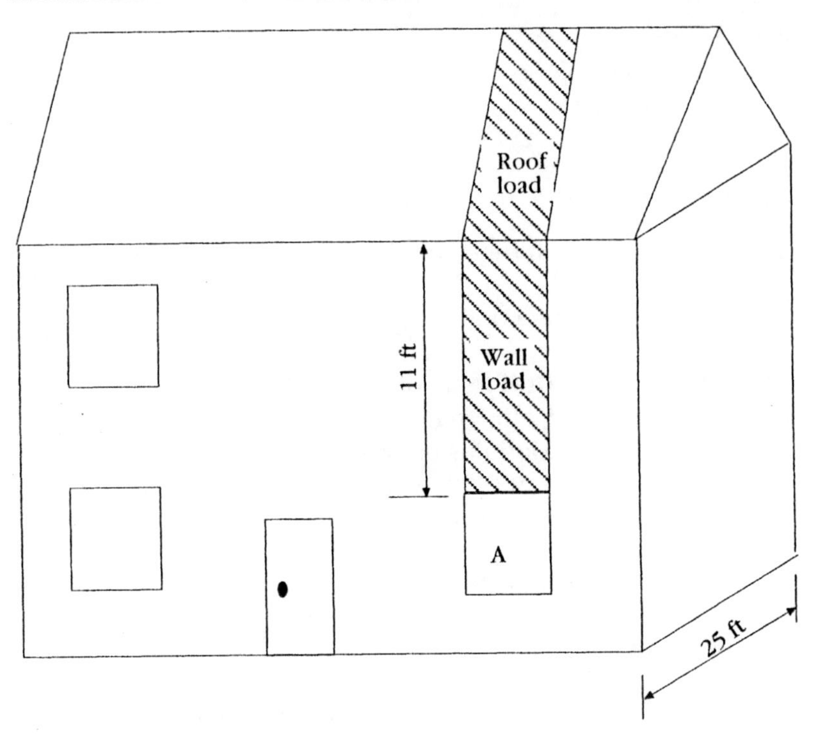

Determine Loads:

Dead loads = Roof structure + Wall + Floor

Roof structure = Assume weight is 20 lb/ft^2
= $20 \times 25/2$
= 250 lb/ft

Wall = Assume weight is 10 lb/ft^2
= 10×11
= 110 lb/ft

Floor = Assume weight is 10 lb/ft^2
= $10 \times 25/2$
= 125 lb/ft

Dead load = 250 + 110 + 125
= 485 lb/ft

Snow load = $20 \times 25/2$
= 250 lb/ft

Live load = $(30 + 40) \times 25/2 + 250$
= 1125 lb/ft
= 93.8 lb/in.

Total load = 485 + 250 +875
= 1610 lb/ft
= 134 lb/in.

Check Deflection:

The maximum deflection for a simply supported header is:

Maximum deflection = $5w\ell^4/384EI$

where w = Uniformly distributed load
ℓ = Span length
E = Modulus of elasticity
= 1,600,000 psi
I = Moment of inertia
= 2×178
= 356 in.^{4}

Deflection for total load is:

$$\Delta_t = 5 \times 134 \times (\ell)^4/(384 \times 1{,}600{,}000 \times 356)$$

$$\Delta_t = \ell^4/326{,}457{,}313$$

Δ_t is limited to $\ell/240$ for headers used in floors.

$$\ell/240 = \ell^4/326{,}457{,}313$$
$$\ell^3 = 1{,}360{,}239$$
$$\ell = 110 \text{ in.}$$
$$\ell = 9.2 \text{ ft}$$

Deflection for live load is:

$$\Delta_L = 5 \times 93.8 \times (\ell)^4/(384 \times 1{,}600{,}000 \times 356)$$

$$\Delta_L = \ell^4/(466{,}367{,}591)$$

Δ_L is limited to 1/360 for live loads on floors.

$$\ell/360 = \ell^4/(466{,}367{,}594)$$
$$\ell^3 = 1{,}295{,}466$$
$$\ell = 118 \text{ in.} = 109$$
$$\ell = 9.1 \text{ ft}$$

Therefore deflection limitations only allow the span to be 9.1 ft or less in length.

The bending stress caused by the loads is:

$$\text{Bending stress} = Md/2I$$

where M = bending moment
d = depth of joist
= 11.25
I = moment of inertia
= 356 in.^{4}

The maximum bending moment for a simply supported beam is:

$$M = w\ell^2/8$$

Substituting this into the bending stress equation gives:

$$\begin{aligned}\text{Bending stress} &= Md/2I \\ &= w\ell^2 d/(8 \times 2 \times I) \\ &= w\ell^2 d/16I\end{aligned}$$

Therefore the bending stress is:

$$\begin{aligned}\text{Bending stress} &= (134 \times \ell^2 \times 11.25)/(16 \times 356) \\ &= \ell^2/3.8\end{aligned}$$

Assuming an allowable bending stress of 1000 psi, the maximum allowable length is:

$$\begin{aligned}1000 &= \ell^2/3.8 \\ 3800 &= \ell^2 \\ \ell &= 61.6 \text{ in.} \\ &= 5.1 \text{ ft}\end{aligned}$$

The shear stress for a rectangular wood section is calculated as follows:

$$\text{Shear stress} = 1.5V/bd$$

where V = Shear force
b = Width of header
= 2×1.5
= 3 in.
d = depth of header
= 11.25 in.

The shear, V, for a uniformly loaded header is:

$$V = w\ell/2$$

As previously discussed, the loads within distance d of the end of the beam are ignored in shear calculations. This allows a reduction multiplier of:

$$\text{Multiplier} = (1 - 2d/\ell)$$

This gives a V equal to:

$$V = (w\ell)(1 - 2d/\ell)/2$$

Substituting this into the shear equation gives:

$$\text{Shear stress} = 0.75(w\ell)(1 - 2d/\ell)/bd$$

Substituting in the known values yields:

$$\text{Shear stress} = 0.75(134 \times \ell)(1 - (2 \times 11.25/\ell)/(11.25 \times 3)$$

$$= (101\ell)(1 - \frac{22.5}{\ell})/33.8$$

$$= 3\ell - 67$$

Given an allowable shear of 90 psi, the maximum span length is:

$$90 = 3\ell - 67$$

$$3\ell = 90 + 67$$

$$\ell = 52.3 \text{ in.}$$

$$= 4.4 \text{ ft}$$

Check Bearing Stress:

The compression force between the bottom of the header ends and the studs supporting it is:

$$\text{End reaction} = w\ell/2$$

The bearing stress is given by:

$$\text{Bearing stress} = \frac{\text{End reaction}}{\text{Area of support}}$$

If the allowable bearing stress equals 600 psi, the area of support needed is:

$$600 = \frac{w\ell/2}{\text{Area of support}}$$

Assuming the header is supported by two 2 × 4s, it will give a maximum length of:

$$600 = \frac{(134 \times \ell)/2}{(2 \times 1.5 \times 3.0)}$$

$$\ell = 80.6 \text{ in.}$$
$$= 6.7 \text{ ft}$$

In summary, the maximum length calculations are:

For dead load deflections	$\ell = 9.2$ ft
For live load deflections	$\ell = 9.1$ ft
For bending stresses	$\ell = 5.1$ ft
For shear stresses	$\ell = 4.4$ ft ←Controls
For bearing stresses	$\ell = 6.7$ ft

Use a maximum span of 4 ft for this load condition.

Advanced Discussion

Typical wood members have shortcomings that preclude their use for very long spans and heavy loaded, short spans. Some typical shortcomings include:

- Potentially large, centralized defects such as knots and splits
- Modulus of elasticity that will allow larger-than-allowed deflections
- Limited sizes based on tree availability
- Defects in areas of high stresses

The shortcomings of typical wood members can be overcome by using glue laminated wood members. These fabricated wood members consist of thin, long pieces of wood glued together to form a layered member. A glued laminated member is shown in Figure 14-5.

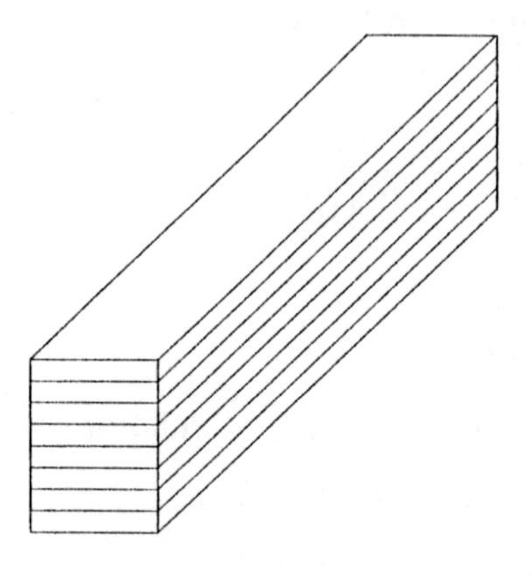

Figure 14-5—Glued laminated beam

When this layered method is used, the following problems have reduced impacts:

- Defects—Since the members are fabricated, each layer can be chosen to have minimal defects. Also, since the members are thin, defects are easily seen and can be dispersed throughout the member, thereby reducing their impact.
- Increased modulus of elasticity—As a result of the overall better-quality product, the modulus of elasticity is larger, and smaller deflections will result.
- The size of the member is only limited by transportation and installation considerations rather than the size of trees.
- Any defects in the laminations can be located at areas of lower stress.

Because of these qualities, glue laminated beams can be efficiently used for headers.

It is interesting to note that the glue that connects the lamentations is stronger than the wood itself. Failure will occur in the wood before it would fail at a glue line.

Glued laminated members have size designations different than that for typical lumber. A 2 × 12 piece of lumber is not 2 in. × 12 in. deep. However, a glued laminated beam is designated by its actual size. A 3.5 × 11 in. glue laminated beam would be those actual dimensions.

The strength properties for glue laminated members are directly related to the quality of the wood laminations that compose the beam. A very rough idea of the changes in proportions from typical lumber is:

Bending stress capacity–Twice as large

Shear stress capacity–Twice as large

Bearing stress capacity–Some increase

Deflection reduction–Some increase

Therefore, when a glue laminated wood is used, the beam can carry more load with less deflection.

Example 2

Assume the same facts as the previous header design example, except use a 3.5 × 11 glue laminated beam for the header. Assume the following properties for the glue laminated beam as compared to the previous example.

Allowable bending stress = Twice as large

Allowable shear stress = Twice as large

Allowable bearing stress = 10% larger

Modulus of elasticity = 10% larger

Note that the glued laminated beam is 3.5-in. thick × 11-in. deep, as opposed to the two regular headers, which are a total of 3 in. (2 × 1.5 = 3) thick × 11.25-in. deep. This will change the geometric properties as follows.

$$\text{Inertia} = bd^3/12$$
$$= 3.5 \times (11)^3/12$$

= 388 (compared to 356 for the wood header)
= 388/356 = 9% increase

Beam width = 3.5 in. (compared to 2 × 1.5 = 3.0)

$$= \frac{3.5}{3.0} = 17\% \text{ increase}$$

Bearing area = (3.5 × 3.0) = 10.5
As compared to (2 × 1.5 × 1.5 × 2 = 9)

$$= \frac{10.5}{9} = 17\% \text{ increase}$$

Deflections:

The allowable span length can be increased since the modulus of elasticity is larger and the moment of inertia is increased. Since the deflection limitation equation is a function of the fourth power of the length, the modifier is the product of the percentage increases:

$$\text{Deflection modifier} = [(10\% \text{ increase}) \times (9\% \text{ increase})]^{1/4} = (1.10 \times 1.09)^{1/4} = 1.05$$

Bending Stress:

The allowable span length can be increased since the allowable bending stress has doubled and the moment of inertia has increased. Since the bending stress limitation equation is a function of the square of the length, the modifier is the product of these increases.

$$\text{Bending stress modifier} = [(\text{double}) \times (9\% \text{ increase})]^{1/2} = (2.0 \times 1.09)^{1/2} = 1.48$$

Shear Stress:

The allowable span length can be increased since the allowable shear stress has doubled and the beam width has increased 16 percent. For the case of computation, ignore the 16% increase.

From the previous example, the equation for allowable span length for shear is modified as follows:

$$(90(2.0) = 3\ell - 67$$
$$3\ell = 180 + 67$$
$$\ell = 82 \text{ in.}$$
$$= 6.9 \text{ ft}$$

Use 7.0 because the increase for larger beam width was ignored.

Bearing Stress:

Assuming the allowable bearing is 10% greater than the wood header, the modifier, including the additional increase caused by the increased bearing area, is:

$$\text{Bearing stress modifier} = 1.17 \text{ (17\% increase)} \times 1.10 = 1.29$$

Therefore the beam lengths for a glued laminated beam are:

Dead load deflections $= 9.2 \times$ Deflection modifier
$= 9.2 \times 1.05$
$= 9.7$ ft

Live load deflections $= 9.1 \times$ Deflection modifier
$= 9.1 \times 1.05$
$= 10.0$ ft

Bending stress $= 5.1 \times$ Bending-stress modifier
$= 5.1 \times 1.48$
$= 7.5$ ft

Shear stress $= 7.0$ ←Controls

Bearing stress $= 6.7 \times$ Bearing-stress modifier
$= 6.7 \times 1.29$
$= 8.6$ ft

By substituting a glued laminated beam of the approximate same size as the two 2 × 12s the span length can be increased from 4 ft to 7 ft.

Steel Members

Wood members are utilized throughout the residential structure because they are adaptable to a wide variety of situations. However, wood cannot always be used as a result of its limited strength. In these circumstances steel becomes the material of choice.

The next two chapters provide information on the design of steel members. Chapter 15 provides information on the design of steel beams. Chapter 16 provides the necessary background for selection of the steel members used to support the brick above an opening in the wall.

15

Steel Beams

Introduction

Steel beams are utilized below the first floor level and in garages to reduce the span length of the wood joists. It would be impractical and maybe impossible for the wood floor joists to span the entire length of the basement (e.g., wall to wall). The steel beam provides support to the wood joists so that the span lengths can be typically reduced to one-half the basement width. Steel beams might also be required if there is some unusually heavy load that would exceed the capacity of a wood member.

The .design of a steel beam requires that the following items be investigated for conformance with the applicable building code.

1. Bending stress
2. Shear stress
3. Deflection

Based on the limitations provided by these three items, Table 15-1 was prepared to facilitate an expedient and efficient method for proper steel beam selection. The calculations necessary to produce Table 15-1 are explored in detail the Advanced Discussion.

Note that the preparation of Table 15-1 was achieved by making various simplifying assumptions that can be made when doing residential design. These assumptions, and therefore Table 15-1

Table 15-1—Steel Beam Distributed Load Capacities (lb/ft)

Size		Length (ft)										
		8	9	10	11	12	13	14	15	16	17	18
W10 × 26	Total	6905	5456	4419	3652	3069	2615	2254	1964	1726	1529	1364
	Live	6905	5456	4419	3652	3069	2615	2254	1833	1510	1259	1060
W10 × 22	Total	5741	4536	3674	3036	2552	2174	1874	1633	1435	1271	1134
	Live	5741	4536	3674	3036	2552	2174	1874	1502	1237	1031	869
W10 × 19	Total	4652	3676	2977	2460	2068	1762	1518	1323	1163	1030	919
	Live	4652	3676	2977	2460	2068	1762	1507	1225	1010	842	709
W10 × 17	Total	4009	3168	2565	2120	1782	1518	1308	1140	1002	887	792
	Live	4009	3168	2565	2120	1782	1518	1308	1031	849	708	596
W8 × 24	Total	5172	4087	3310	2735	2299	1958	1688	1471	1292	1145	1021
	Live	5172	4087	3310	2673	2059	1619	1296	1054	868	724	610

Table 15-1—Steel Beam Distributed Load Capacities (lb/ft) (cont.)

W8 × 21	Total	4503	3559	2882	2381	2002	1705	1469	1280	1125	997	889
	Live	4503	3559	2882	2381	1872	1473	1179	959	790	658	555
W8 × 18	Total	3761	2972	2407	1989	1672	1424	1227	1069	940	833	743
	Live	3761	2972	2407	1989	1539	1210	969	787	649	541	456
W8 × 15	Total	2920	2307	1868	1544	1298	1105	953	830	729	646	576
	Live	2920	2307	1868	1544	1193	939	751	611	503	419	353
W8 × 13	Total	2452	1937	1569	1297	1090	928	800	697	613	543	484
	Live	2452	1937	1569	1278	984	774	620	504	415	346	291

and the following discussion is valid only for residential structures. In particular, it is only for uniform distributed loading for steel beams of ASTM A36 Steel.

Steel beams used in residential construction are most likely one of the sizes shown in Table 15-2. Table 15-2 also contains the pertinent properties for each of these beams.

The appropriate steel beam size cannot be determined unless the applied loads are known. The potential loads on a steel beam can be classified into the following groups:

1. Dead load (weight of structure)
2. Live load (load applied by occupants)
3. Snow load
4. Wind load

Table 15-2—Typical Steel Members Used in Residential Construction

Designation	Area (in.^{2})	Depth (in.)	Web Thickness (in.)	I (in.^{4})	Flange Width (in.)	Flange Thickness (in.)
W10 × 26	7.61	10.33	0.260	144	5.77	0.44
W10 × 22	6.49	10.17	0.240	118	5.75	0.36
W10 × 19	5.62	10.24	0.250	96	4.02	0.39
W10 × 17	4.99	10.11	0.240	82	4.01	0.33
W8 × 24	7.08	7.93	0.245	83	6.50	0.40
W8 × 21	6.16	8.28	0.250	75	5.27	0.40
W8 × 18	5.26	8.14	0.230	62	5.25	0.33
W8 × 15	4.44	8.11	0.245	48	4.01	0.31
W8 × 13	3.84	7.99	0.230	40	4.00	0.25

Figures 15-1 and 15-2, respectively, provide a plan view and a section of a residential structure. The steel beam to be designed is denoted as steel beam "A" in these figures. The potential loading conditions for this steel beam will be discussed individually.

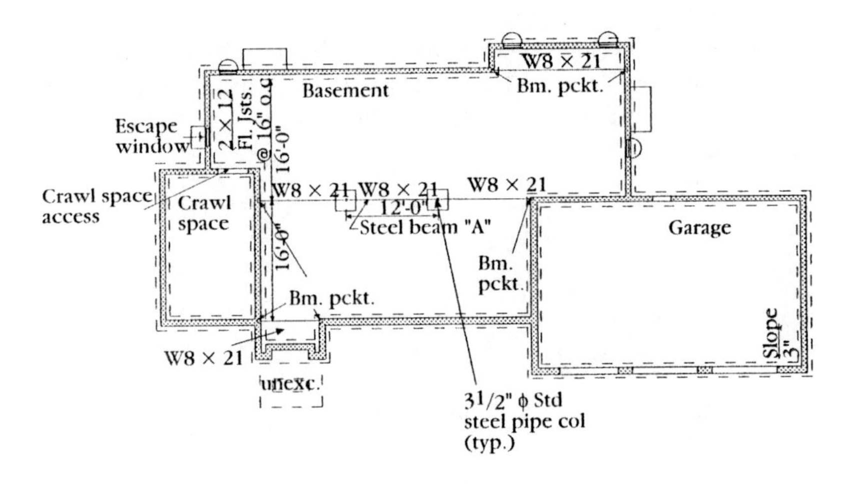

Figure 15-1—Plan view of foundation level of a residential structure

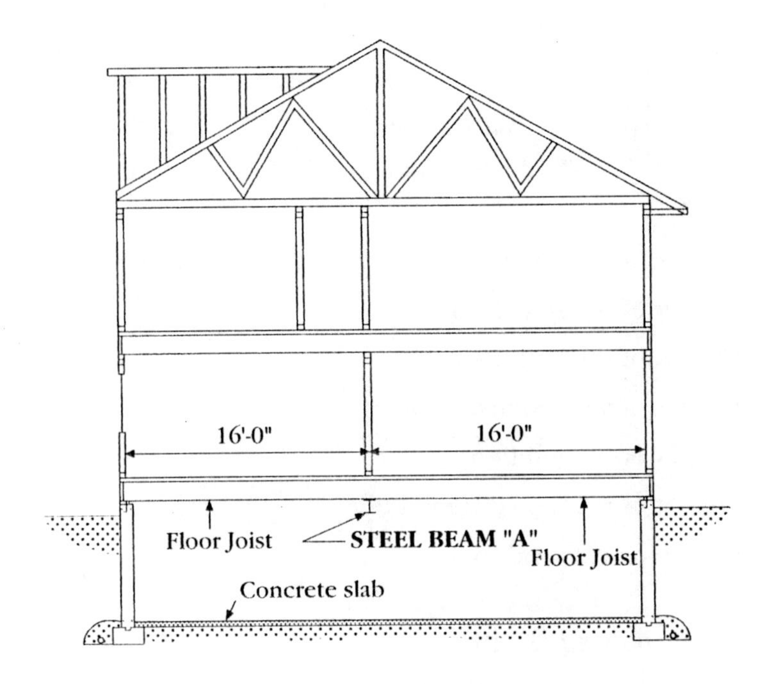

Figure 15-2—Section of residential structure

1. Dead Load

The tributary area that will transmit dead load to the steel beam is shown in Figure 15-3. The dead load (i.e., the weight of the structure) that must be carried by the steel beam includes the following:

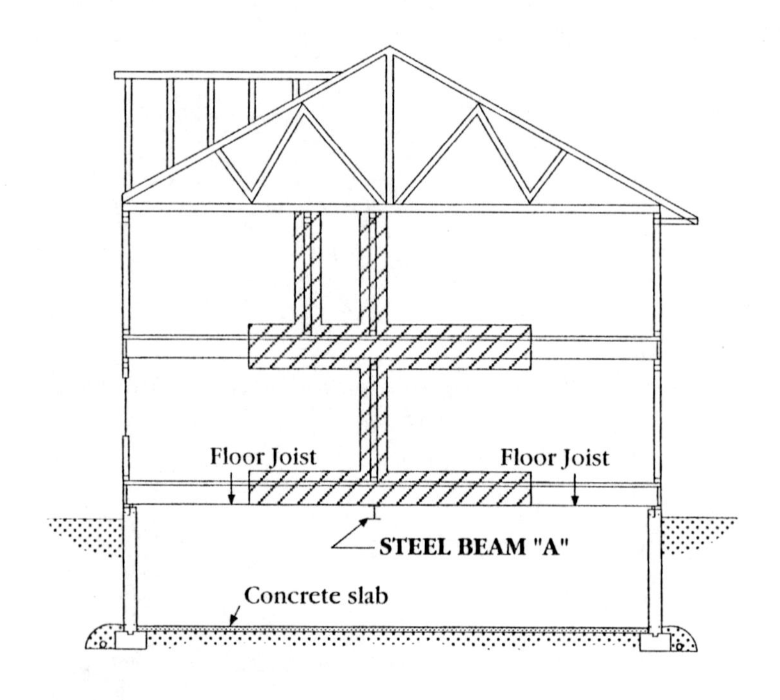

Figure 15-3—Dead load that is transferred to steel beam "A"

Weight of the steel beam

Weight of the first floor

Weight of wall on first floor

Weight of second floor

Weight of walls on second floor

Weight of attic and roof

Weight of Steel Beam

In addition the loads that are imposed on the steel beam, the beam must also carry its own weight. The weight of a steel beam used in residential construction would typically range from 15 to 25 lb/ft. A beam weight of 21 lb/ft is used as an estimate for this beam design. Note that beam weight is only a small percentage of the overall loads carried by the beam and an incorrect first guess will probably not change the results.

Weight of the First Floor

The weight of the first floor includes a floor covering, plywood decking, joists and possibly drywall on the underside of the floor if the basement is a finished room. These items produce a load of 10 lb/ft^2. Therefore, the load carried by the steel beam from the first floor weight is:

$$\text{Load} = 10 \times 16 \text{ ft} = 160 \text{ lb/ft}$$

Weight of Wall on First Floor

There is a wall located directly above the steel beam. The weight of this wall consists of the weight of the wood studs and plates, drywall on both sides of the wall and any item enclosed in the wall. It is estimated that this amounts to 10 lb/ft^2. If the wall is estimated to be 8-ft high, the load on the steel beam is:

$$\text{Load} = 10 \times 8 \text{ ft} = 80 \text{ lb/ft}$$

Weight of Second Floor

The second-floor weight is considered the same as that of the first floor, which is 160 lb/ft.

Weight of Walls on Second Floor

There are two walls on the second floor that must be supported by the steel beam. Using 10 lb/ft^2 and an 8-ft wall height, the load imposed by the two walls is:

$$\text{Load} = 2 \times 10 \times 8 \text{ ft} = 160 \text{ lb/ft}$$

Weight of Attic and Roof

Since the roof consists of trusses that span the entire width of the structure, the steel beam does not support any of the attic or roof loads.

The total dead load that must be carried by the steel beam is therefore:

Weight of steel beam	21 lb/ft
Weight of first floor	160 lb/ft
Weight of wall on first floor	80 lb/ft
Weight of second floor	160 lb/ft
Weight of walls on second floor	160 lb/ft
Weight of attic and roof	0 lb/ft

Total dead load = 581 lb/ft

2. Live Load

The tributary area that will transmit live loads to the steel beam is shown in Figure 15-4. The live load (i.e., floor loads as a result of use) that must be carried by the steel beam includes the following:

Live load on first floor

Live load on second floor

Attic and roof load

Live Load on First Floor

The first floor, a nonsleeping area, is required to be designed for a 40-lb/ft^2 live load. The live load that must be carried in the tributary area of the steel beam produces the load shown in the following calculation:

$$\text{Load} = 40 \times 16 \text{ ft} = 640 \text{ lb/ft}$$

Live Load on Second Floor

The second floor is a sleeping area and is required to be designed for a 30-lb/ft^2 live load. The live load that must be carried in the

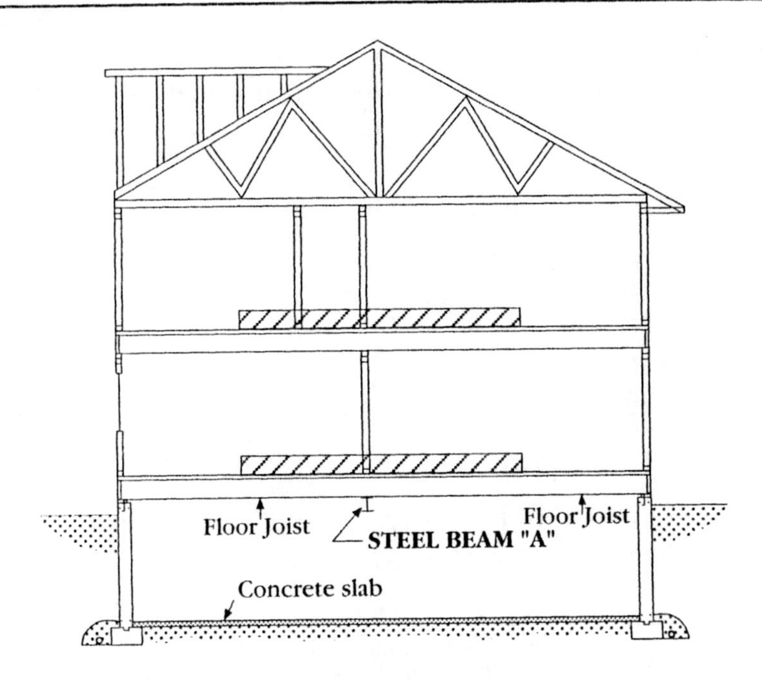

Figure 15-4–Live load that is transferred to steel beam "A"

tributary area of the steel beam produces the load shown in the following calculation:

Load = 30 × 16 ft = 480 lb/ft

Attic and Roof Load

Since the roof is constructed of trusses that span the width of the structure, the steel beam does not support any attic or roof loads.

The total live load that must be carried by the steel beam is:

Live load on first floor	640 lb/ft
Live load on second floor	480 lb/ft
Attic and roof load	0 lb/ft

Total live load = 1120 lb/ft

Snow Loads and Wind Loads

Since the roof is constructed of trusses that span the width of the structure, the steel beam does not support any snow or wind loads.

Therefore, the load on the steel beam is:

Dead load	581 lb/ft
Live load	1120 lb/ft
Snow load	0
Wind load	0

Total load = 1701 lb/ft

Referring to Table 15-1, for a 12-ft span, the maximum loads for each of the presented beams are:

Size	Maximum Allowable Total Load	Maximum Allowable Live Load
W10 × 26	3069	3069
W10 × 22	2552	2552
W10 × 19	2068	2068
W10 × 17	1782	1782
W8 × 24	2299	2059
W8 × 21	2002	1872
W8 × 18	1672	1539
W8 × 15	1298	1193
W8 × 13	1090	985

Review of the maximum allowable live load shows that with the exception of the W8 × 13, all beams would be satisfactory. Review of the maximum allowable total load shows that both the W8 × 15 and W8 × 18 are also unacceptable.

The most cost-effective choice would be the lightest beam since steel prices are based on weight. From this point of view, the W10 × 17 would be the best choice since it weighs only 17 lb/ft. However, the W8 × 21 is only 4 lb heavier but is 2 in. smaller in depth. The extra headroom in the basement might be worth the extra cost for the steel beam.

Advanced Discussion

This section provides the details for the design of the steel beams presented in the introductory portion of this topic. The discussion herein applies only to residential and comparable light construction.

In the introductory portion of this topic, it was determined that the uniformly distributed loads on the beam were:

Live load	1120 lb/ft
Dead load	581 lb/ft
Snow load	0
Wind load	0
Total load	1701 lb/ft

As previously discussed, the maximum bending stress for a uniformly distributed load on a simply supported beam is:

Maximum bending moment = $w\ell^2/8$

where w = total uniformly distributed load
= 1701 lb/ft (142 lb/in.)
ℓ = span length of beam
= 12 ft (144 in.)

Substituting these numbers into the equation gives the maximum bending moment of:

Maximum bending moment = $w\ell^2/8$
= $(142)(144)^2/8$
= 368,064 lb/in.

The bending stress produced by this moment is determined by:

Bending stress = Md/2I

where M = Maximum bending moment
= 368,064 lb/in.
d = Depth of steel beam
I = Moment of Inertia of steel beam

For a W8 × 21 the depth, d, is 8.28 in. and the moment of inertia, I, is 75 in.^{4} (see Table 15-2). Substituting in these numbers the value of the bending stress is:

$$\text{Bending stress} = \frac{(368{,}064)(8.82)}{(2)(75)}$$

$$= 20{,}317 \text{ lb/in.}^2$$

The maximum allowable bending stress for ASTM A36 steel under these specific loading conditions is two-thirds of the yield stress (2/3 × 36) = 24,000 psi. Because 24,000 psi > 20,317 psi, the steel satisfies bending moment requirements.

As previously discussed, the maximum shear stress for a simple supported beam with a uniform load is:

$$\text{Maximum shear} = w\ell/2$$

where w = Total uniformly distributed load
= 1701 lb/ft (142 lb/in.)
ℓ = Length of steel beam
= 12 ft (144 in.)

Substituting these numbers into the equation, the maximum shear is:

$$\begin{aligned}\text{Maximum shear} &= w\ell/2\\ &= (142)(144)/2\\ &= 10{,}224 \text{ lb}\end{aligned}$$

The maximum shear stress, f, is determined using this shear of 10,224 lb:

$$f = V/(dt_w)$$

where V = Maximum shear
= 10,224 lb

d = Depth of steel beam
= 8.28 in.

t_w = Web thickness
= 0.250 in. for a W8 × 21

Using these values the shear stress is:

$$f = 10{,}224/(8.28 \times 0.250)$$
$$= 4939.1 \text{ lb/in.}^2$$

The maximum allowable shear stress for ASTM A36 steel is 14,400 lb/in.^{2} (0.4 × 36,000 psi = 14,400 psi). This steel member easily satisfies shear stress criteria since the allowable shear stress of 14,400 psi is greater than the applied shear stress of 4939.1 lb/in.^{2}.

As discussed previously, the maximum deflection for a simple supported beam is given by:

$$\text{Maximum deflection} = 5w\ell^4/(384EI)$$

where: w = Uniform applied live load
= 1120 lb/ft (93.3 lb/in.)

ℓ = Length of steel beam
= 12 ft (144 in.)

E = Modulus of elasticity for steel
= 29,000,000 lb/in.^{2}

I = Moment of inertia
= 75 in.^{4}

Substituting these numbers into the equation, the maximum deflection is:

$$\text{Maximum deflection} = \frac{5 \times 93.3 \times (144)^4}{384 \times 29{,}000 \times 75}$$

$$= 0.24 \text{ in.}$$

The maximum allowable deflection caused by live loads cannot exceed the span length divided by 360. Therefore, the maximum allowable deflection is:

$$\text{Maximum allowable deflection} = \ell/360 = 144/360 = 0.4 \text{ in.}$$

Deflection criteria is satisfied since the allowable deflection of 0.4 in. is greater than the maximum deflection caused by the loads, 0.24 in.

The following summarizes the design of the steel beam.

	Maximum	Allowable	Percent Utilized	Conclusion
Bending Stress	20,317	24,000	85	Satisfied
Shear Stress	4939	14,400	34	Satisfied
Deflection	0.24	0.40	60	Satisfied

Review of this table shows that for this beam the bending stresses control the design since 85% of capacity is utilized.

These calculations can also be verified using Table 15-1 from the Introduction section of this topic. For a 12-ft-long W8 × 21 the comparison is:

	As Calculated	From Table 1	Percent Capacity Utilized
Live Load	1120	1872	60
Total Load	1701	2002	85

The 85% of capacity previously discussed verifies the previous calculations.

16

Steel Lintels

Introduction

A functional house must have openings in the structure to provide exits, ventilation and light. These windows and doors result in discontinuities in the wood framing and exterior cladding of the house. The disruption in the continuity produces no detrimental changes in the loads when the exterior cladding is wood, vinyl or aluminum siding. This is not the case when the exterior is brick.

The load produced by the weight of the brick exterior is transferred down from brick to brick to the foundation. When openings for window and doors are present in the brick exterior, the brick in the area above the window cannot transfer its weight directly down. For instance, if an opening is placed in a brick exterior for a window, the brick directly above the window supports the weight of a portion of the brick above. The combination of the weight of the brick above the window opening and lack of support for the layer of bricks directly below will cause the bricks to deflect downward. This downward movement will crack the wall or even collapse the brick exterior.

It is apparent then that some type of structural support is needed at the top of openings to prevent downward movement of the brick directly above the openings. This structural support is called a *lintel*. A lintel is a structural member that will carry the load over an opening to the adjacent sides. The load is then carried down from the sides of the opening to the foundation.

An ideal structural member for a lintel is the steel angle. As shown in Figure 16-1 the brick sits directly on one of the legs of the angle. Although not very common in residential construction, lintels can also be constructed of reinforced concrete or masonry.

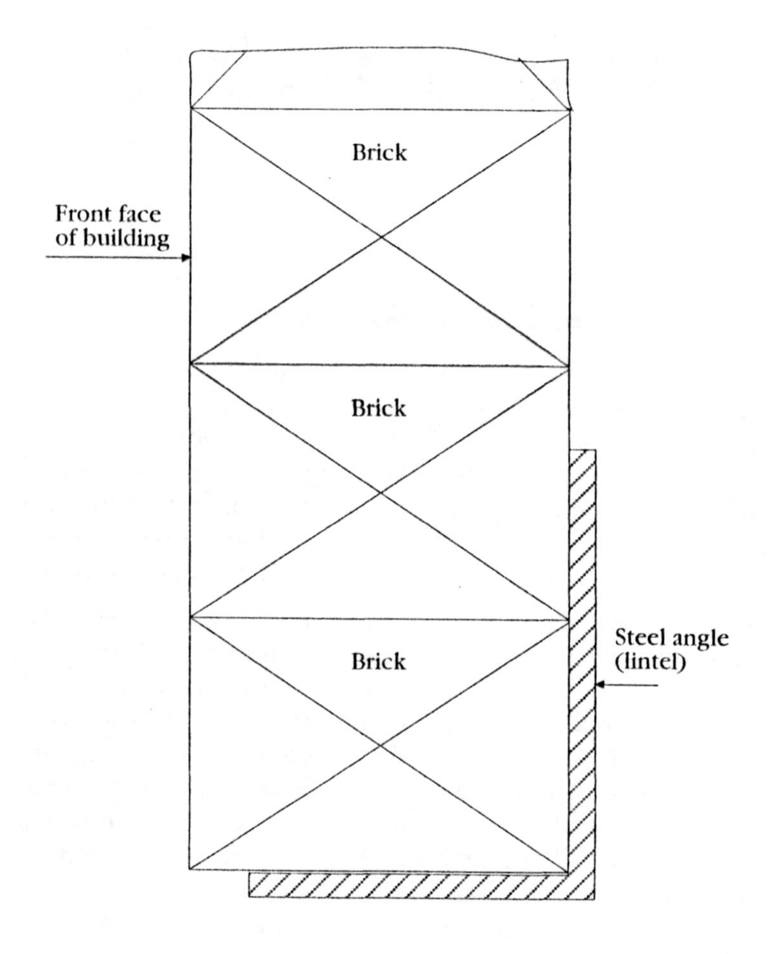

Figure 16-1—Lintel supporting brick

Notice that the brick wall is connected to the wood framing with steel straps. This connection is merely to prevent the brick from moving away from the face of the building. The steel straps are not stiff enough to transfer any loads. For example, the snow loads on the roof are carried by the roof plywood to the roof trusses. The load is transferred from the roof trusses to the exterior wood walls and then down to the foundation. The brick exterior does not participate in the support or load transfer of the snow load. Residential brick exterior walls used for cladding only carry the load of their own weight. If the brick walls carry external loads other than their weight, this section is inapplicable and it is wise to retain a structural engineer to do the calculations.

For corrosion and other practical purposes, 1/4 in. or greater is used for the thickness of the steel angle. Also the bottom leg of the angle must be of a sufficient length for the brick to have support. The bottom leg must not be too long or it will stick out past the face of the brick. For this reason an angle must have the bottom leg dimension of 3-1/2 or 4 in. Based on the minimum thickness, minimum and maximum length of the bottom leg and availability of sizes, Table 16-1 lists the steel angles that can be used as lintels and the corresponding properties.

Lintels must be strong enough to carry the load of the bricks as well as have an acceptable amount of deflection. To reach this goal, the stress is limited to no more than 24,000 psi or less, depending on the dimensions of the angle, and deflections are limited to the length of the opening divided by 600 (that is, $\ell/600$). Tables 16-2 and 16-3 present the maximum height of brick that can be supported by the angle for the given span length. Table 16-2 is for span lengths up to 8 ft. Table 16-3 is for span lengths greater than 8 ft but 14 ft or less. Assumptions used in preparing these results were:

- Lintel only supports the exterior brick
- Weight of brick is 120 lb/ft^3
- The benefit of brick arching was not taken into account

Example–A steel angle is needed to support 3 ft of brick located over an 8-ft-wide opening. What size angle should be used?

Referring to Table 16-2, a 3-1/2 × 3-1/2 × 3/8 can support 3.6 ft high of brick. A 4 × 3-1/2 × 1/4 is capable of carrying 3.7 ft high of brick. Although both angles can support the 3 ft of brick, the 4 × 3-1/2 × 1/4 is more than 2 lb/ft lighter than the 3-1/2 × 3-1/2 × 3/8 angle. Therefore it will be cheaper as well as easier to install.

Table 16-1—Steel Angle Geometric Properties

Angle Designation	Vertical Leg (in.)	Horizontal Leg (in.)	Thickness (in.)	Weight per Foot (pounds)	I (in.^{4})	S (in.^{3})
3 × 3-1/2 × 1/4	3	3-1/2	1/4	5.4	1.30	0.59
3 × 3-1/2 × 5/16	3	3-1/2	5/16	6.6	1.58	0.72
3 × 3-1/2 × 3/8	3	3-1/2	3/8	7.9	1.85	0.85
3-1/2 × 3-1/2 × 1/4	3-1/2	3-1/2	1/4	5.8	2.01	0.79
3-1/2 × 3-1/2 × 5/16	3-1/2	3-1/2	5/16	7.2	2.45	0.98
3-1/2 × 3-1/2 × 3/8	3-1/2	3-1/2	3/8	8.5	2.87	1.15
4 × 3-1/2 × 1/4	4	3-1/2	1/4	6.2	2.91	1.03
4 × 3-1/2 × 5/16	4	3-1/2	5/16	7.7	3.56	1.26
4 × 4 × 1/4	4	4	1/4	6.6	3.04	1.05
4 × 4 × 5/16	4	4	5/16	8.2	3.71	1.29
4 × 4 × 3/8	4	4	3/8	9.8	4.36	1.52
4 × 4 × 1/2	4	4	1/2	12.8	5.56	1.97

Table 16-1—Steel Angle Geometric Properties (cont.)

5 × 3-1/2 × 5/16	5	3-1/2	5/16	8.7	6.60	1.94
5 × 3-1/2 × 3/8	5	3-1/2	3/8	10.4	7.78	2.29
5 × 3-1/2 × 1/2	5	3-1/2	1/2	13.6	9.99	2.99
6 × 3-1/2 × 5/16	6	3-1/2	5/16	9.8	10.9	2.73
6 × 3-1/2 × 3/8	6	3-1/2	3/8	11.7	12.9	3.24
6 × 4 × 3/8	6	4	3/8	12.3	13.5	3.32
6 × 4 × 1/2	6	4	1/2	16.2	17.4	4.33

Table 16-2—Allowable Height of Brick That Can Be Supported by Steel Angle—Short Spans

	Span ft					
Angle Size	3	4	5	6	7	8
3 × 3-1/2 × 1/4	*	13.0	6.7	3.9	2.4	1.6
3 × 3-1/2 × 5/16	*	15.9	8.1	4.7	3.0	2.0
3 × 3-1/2 × 3/8	*	*	9.5	5.5	3.5	2.3
3-1/2 × 3-1/2 × 1/4	*	*	10.4	6.0	3.8	2.5
3-1/2 × 3-1/2 × 5/16	*	*	12.6	7.3	4.6	3.1
3-1/2 × 3-1/2 × 3/8	*	*	14.8	8.6	5.4	3.6
4 × 3-1/2 × 1/4	*	*	15.0	8.7	5.5	3.7
4 × 3-1/2 × 5/16	*	*	*	10.6	6.7	4.5
4 × 4 × 1/4	*	*	15.7	9.1	5.7	3.8
4 × 4 × 5/16	*	*	*	11.1	7.0	4.7
4 × 4 × 3/8	*	*	*	13.0	8.2	5.5
4 × 4 × 1/2	*	*	*	16.6	10.4	7.0

***Denotes height in excess of 18 ft**

Table 16-3—Allowable Height of Brick That Can Be Supported by Steel Angle—Long Spans

	Span ft					
Angle Size	9	10	11	12	13	14
3-1/2 × 3-1/2 × 3/8	2.5	1.8	1.4	0	0	0
4 × 3-1/2 × 1/4	2.6	1.9	1.4	0	0	0
4 × 3-1/2 × 5/16	3.1	2.3	1.7	1.3	0	0
4 × 4 × 1/4	2.7	2.0	1.5	1.1	0	0
4 × 4 × 5/16	3.3	2.4	1.8	1.4	1.1	0
4 × 4 × 3/8	3.9	2.8	2.1	1.6	1.3	0
4 × 4 × 1/2	4.9	3.6	2.7	2.1	1.6	1.3
5 × 3-1/2 × 5/16	5.8	4.3	3.2	2.5	1.9	1.6
5 × 3-1/2 × 3/8	6.9	5.0	3.8	2.9	2.3	1.8
5 × 3-1/2 × 1/2	8.8	6.4	4.8	3.7	2.9	2.3
6 × 3-1/2 × 5/16	9.6	7.0	5.3	4.1	3.2	2.6
6 × 3-1/2 × 3/8	11.4	8.3	6.2	4.8	3.8	3.0
6 × 4 × 3/8	11.9	8.7	6.5	5.0	4.0	3.2
6 × 4 × 1/2	15.4	11.2	8.4	6.5	5.1	4.1

The bricks that form the wall are bonded together to form a wall that acts as a rigid unit. This monolithic action of the wall can possibly reduce the weight that needs to be carried by the lintel. The brick adjacent to the opening can help carry the weight of the brick above the opening. This is referred to as *arching*.

If the brick walls extend uninterrupted upwards and to the sides, the lintel need only carry the weight of the brick shown in the triangles of Figure 16-2. The weight of the brick above the triangle is not carried by the lintels, rather it is carried by arching of the brick. The arching of the brick will transmit the loads outside the triangle to the sides of the opening.

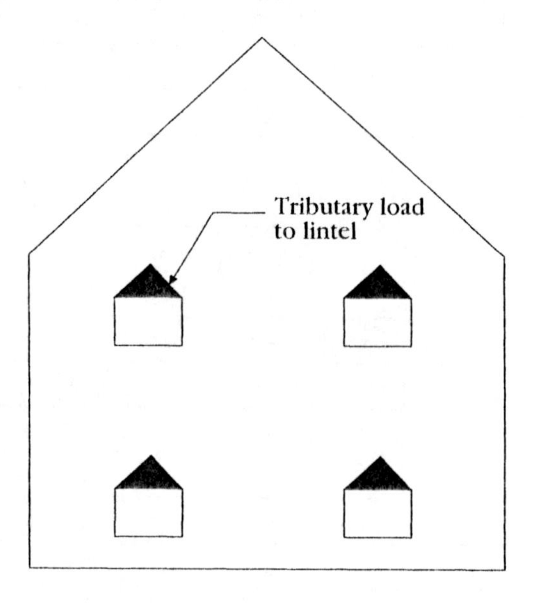

Figure 16-2—Shaded portion depicts the tributary area that is carried by each lintel as the result of arching

The benefit of arching can only be taken advantage of when the brick extends uninterrupted above and to the sides of the opening. It is generally agreed that the brick must extend upward a distance of one-half the opening plus another 1.5 ft. For an 8-ft opening the brick must extend upward a distance of 5.5 ft (8/2 + 1.5 = 5.5).

The length the brick must also extend to the side of the opening to have the benefit of arching action. The uninterrupted length the brick must extend is difficult to determine. A side length of one-half the opening plus an additional 0.5 ft is used in calculations in this text. It is believed that this is a reasonable distance.

Tables 16-4 and 16-5 provide the allowable height of brick that the lintel can carry with the benefit of arching of the brick. Table 16-4 provides results for spans up to 8 ft. Table 16-5 provides results for spans between 8 and 14 ft.

Example–Determine what size angle is needed for the window opening shown in Figure 16-3.

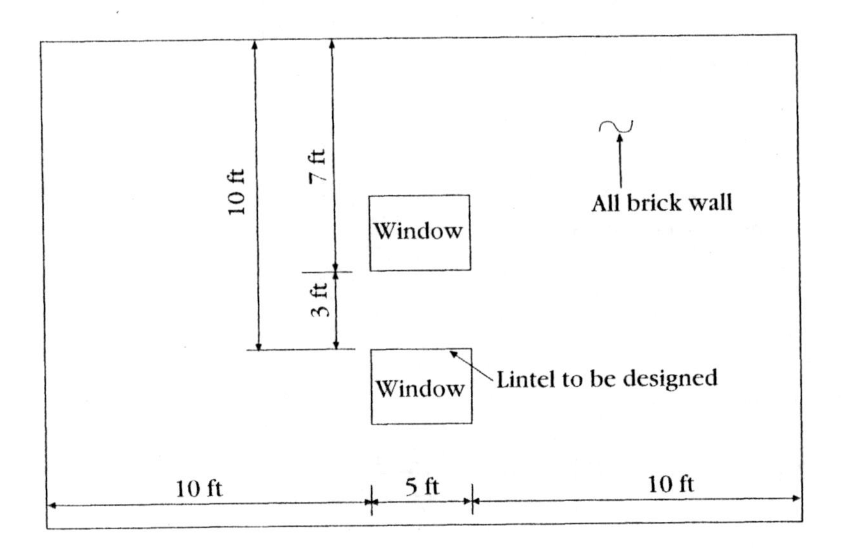

Figure 16-3–Lintel where arching does not occur

The width of the opening is 5 ft. The brick extends uninterrupted upward for 3 ft and horizontally for 10 ft in each direction. Arching will not occur since the uninterrupted brick must be 4 ft above the opening. Therefore, design for 10 ft high of brick. Using Table 16-2, a 3-1/2 × 3-1/2 × 1/4 will be satisfactory.

Example–Determine what size angle is needed for the opening shown in Figure 16-4.

Table 16-4—Special Conditions: Allowable Height of Brick That Can Be Supported by Steel Angle—Short Span

Angle Size	Span ft					
	3 See "A" below	4 See "B" below	5 See "C" below	6 See "D" below	7 See "E" below	8 See "F" below
3 × 3-1/2 × 1/4	No Limit	No Limit	No Limit	No Limit	2.3	1.5
3 × 3-1/2 × 5/16	No Limit	No Limit	No Limit	No Limit	2.8	1.8
3 × 3-1/2 × 3/8	No Limit	No Limit	No Limit	No Limit	No Limit	2.1
3-1/2 × 3-1/2 × 1/4	No Limit	No Limit	No Limit	No Limit	No Limit	2.3
3-1/2 × 3-1/2 × 5/16	No Limit	No Limit	No Limit	No Limit	No Limit	2.9
3-1/2 × 3-1/2 × 3/8	No Limit	No Limit	No Limit	No Limit	No Limit	3.4
4 × 3-1/2 × 1/4	No Limit	No Limit	No Limit	No Limit	No Limit	No Limit

Table 16-4—Special Conditions: Allowable Height of Brick That Can Be Supported by Steel Angle—Short Span (cont.)

4 × 3-1/2 × 5/16	No Limit	No Limit	No Limit	No Limit	No Limit	No Limit
4 × 4 × 1/4	No Limit	No Limit	No Limit	No Limit	No Limit	No Limit
4 × 4 × 5/16	No Limit	No Limit	No Limit	No Limit	No Limit	No Limit
4 × 4 × 3/8	No Limit	No Limit	No Limit	No Limit	No Limit	No Limit
4 × 4 × 1/2	No Limit	No Limit	No Limit	No Limit	No Limit	No Limit

"A"—Numbers are only valid if there is 3 ft of uninterrupted brick above lintel and 2 ft of uninterrupted brick on both sides of lintel

"B"—Numbers are only valid if there is 3-1/2 ft of uninterrupted brick above lintel and 2-1/2 ft of uninterrupted brick on both sides of lintel

"C"—Numbers are only valid if there is 4 ft of uninterrupted brick above lintel and 3 ft of uninterrupted brick on both sides of lintel

"D"—Numbers are only valid if there is 4-1/2 ft of uninterrupted brick above lintel and 3-1/2 ft of uninterrupted brick on both sides of lintel

"E"—Numbers are only valid if there is 5 ft of uninterrupted brick above lintel and 4 ft of uninterrupted brick on both sides of lintel

"F"—Numbers are only valid if there is 5-1/2 ft of uninterrupted brick above lintel and 4-1/2 ft of uninterrupted brick on both sides of lintel

Table 16-5—Special Conditions: Allowable Height of Brick That Can Be Supported by Steel Angle—Long Span

Angle Size	Span ft					
	9 See "A" below	10 See "B" below	11 See "C" below	12 See "D" below	13 See "E" below	14 See "F" below
3-1/2 × 3-1/2 × 3/8	2.3	1.6	1.0	0	0	0
4 × 3-1/2 × 1/4	2.4	1.7	1.2	0	0	0
4 × 3-1/2 × 5/16	2.9	2.1	1.5	1.1	0	0
4 × 4 × 1/4	2.5	1.8	1.3	0	0	0
4 × 4 × 5/16	3.0	2.1	1.5	1.1	0	0
4 × 4 × 3/8	3.6	2.5	1.8	1.3	1.0	0
4 × 4 × 1/2	No Limit	3.2	2.3	1.7	1.3	0
5 × 3-1/2 × 5/16	No Limit	4.0	2.9	2.2	1.7	1.3
5 × 3-1/2 × 3/8	No Limit	No Limit	3.5	2.6	2.0	1.5

Table 16-5—Special Conditions: Allowable Height of Brick That Can Be Supported by Steel Angle—Long Span (cont.)

5 × 3-1/2 × 1/2	No Limit	No Limit	No Limit	3.3	2.5	2.0
6 × 3-1/2 × 5/16	No Limit	No Limit	No Limit	3.8	2.9	2.3
6 × 3-1/2 × 3/8	No Limit	No Limit	No Limit	4.5	3.4	2.7
6 × 4 × 3/8	No Limit	No Limit	No Limit	4.7	3.5	2.8
6 × 4 × 1/2	No Limit	No Limit	No Limit	No Limit	4.6	3.6

"A"—Numbers are only valid if there is 6 ft of uninterrupted brick above lintel and 5 ft of uninterrupted brick on both sides of lintel

"B"—Numbers are only valid if there is 6-1/2 ft of uninterrupted brick above lintel and 5-1/2 ft of uninterrupted brick on both sides of lintel

"C"—Numbers are only valid if there is 7 ft of uninterrupted brick above lintel and 6 ft of uninterrupted brick on both sides of lintel

"D"—Numbers are only valid if there is 7-1/2 ft of uninterrupted brick above lintel and 6-1/2 ft of uninterrupted brick on both sides of lintel

"E"—Numbers are only valid if there is 8 ft of uninterrupted brick above lintel and 7 ft of uninterrupted brick on both sides of lintel

"F"—Numbers are only valid if there is 8-1/2 ft of uninterrupted brick above lintel and 7-1/2 ft of uninterrupted brick on both sides of lintel

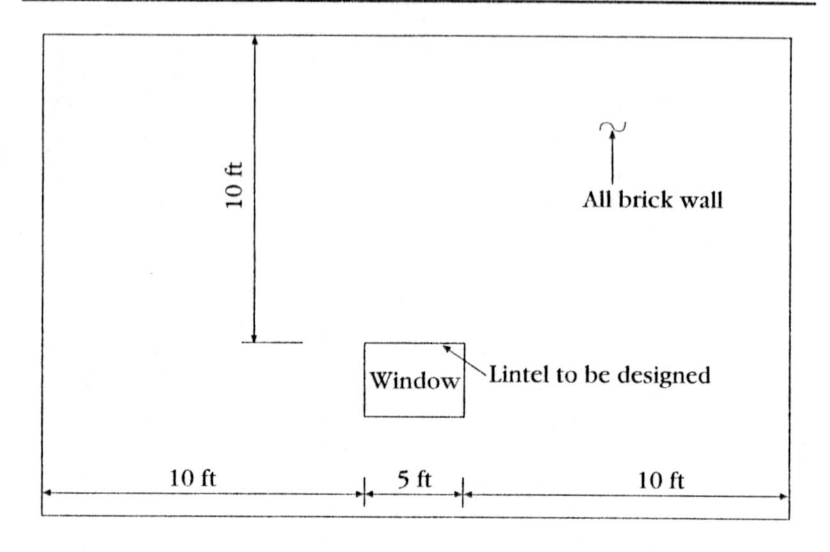

Figure 16-4—Lintel where arching does occur

This is the same as the previous example except that the uninterrupted height is 10 ft. Therefore, arching will occur and Table 16-4 can be used. Referring to this table, a 3 × 3-1/2 × 1/4 will be sufficient. Comparing this to the previous example, arching allowed for a reduction of approximately 1/2 lb/ft.

The ends of the lintels must extend beyond the edge of the opening to properly transfer the weight of the brick to the side of the opening. The length that the angle must extend beyond the opening is a function of the strength of the brick wall, the size of the angle and the weight of the brick carried by the angle.

The following rule can be used for determining the length the angle that should extend beyond the opening

3-or 4-ft opening	4 in.
5-ft opening	5 in.
6-ft or greater	6 in.

Example—Determine the length of angle that should be ordered for the lintel in the previous example.

For a 5-ft opening a 5-in. support length should be provided. Therefore the length of the angle should be:

Length = 5 ft-0 in. + 5 in. + 5 in. = 5 ft-10 in.

Advanced Discussion

This section provides detailed information on the design of a steel lintel. The design of the lintel will include satisfying the requirements for:

1. Bending stresses

2. Deflections

3. Shear stresses

4. Bearing stresses

Note that the discussion in this section applies to lintels that support exterior brick only. If other loads are transferred to the lintel, appropriate modifications must be made.

The lintel to be designed is shown in Figure 16-5. For calculation of the load on the lintel, it will be assumed that the weight of the window is the same as that of the brick. This is obviously conservative but should have minimal impact on the final results.

The weight of the wall on the lintel is:

Height of brick = 12 ft

Weight of brick = 40 lb/ft^2

Load on angle = Height × Weight
= 12 × 40
= 480 lb/ft

The maximum bending stress for a uniformly distributed load on a simply supported beam is:

$$\text{Maximum bending moment} = w\ell^2/8$$

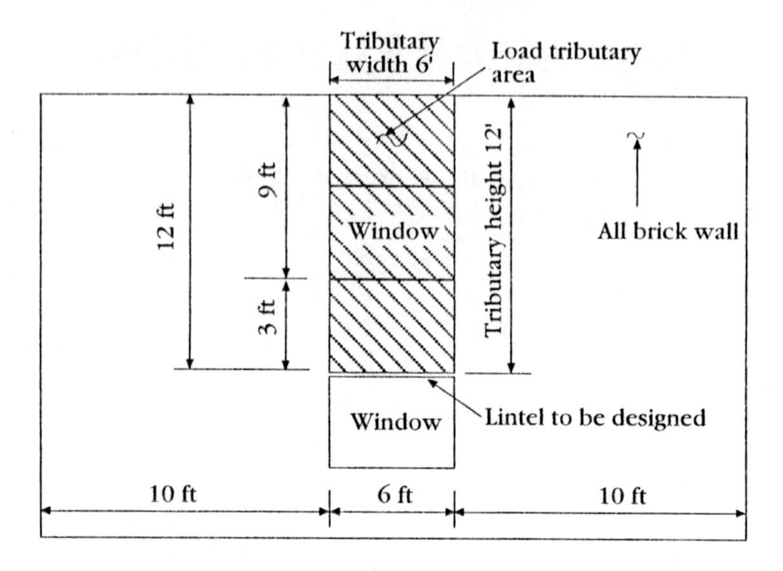

Figure 16-5—Lintel where arching does not occur

where w = Load on angle

= Brick load and weight of steel

= 480 + 10 (assume 10) = 490 lb/ft

= 40.8 lb/in.

ℓ = Width of opening

= 6 ft (72 in.)

Substituting these numbers into the equation gives the maximum bending moment of:

$$\text{Maximum bending moment} = w\ell^2/8 = (40.8)(72)^2/8 = 26{,}438 \text{ in.-lb}$$

The bending stress produced by this moment is determined by:

Bending stress = M/S

where M = Maximum bending moment
= 26,438 in.-lb
S = 1.52 in.^{3} (assume $4 \times 4 \times 3/8$)

Substituting these numbers into the equation gives a bending stress of:

$$\text{Bending stress} = 26{,}438/1.52 = 17{,}393 \text{ lb/in.}^2$$

The maximum allowable bending stress for ASTM A36 steel under these specific loading conditions is two-thirds of the yield stress ($2/3 \times 36{,}000$ psi) = 24,000 psi. Since 24,000 psi > 17,393 psi, the lintel satisfies bending moment requirements.

The maximum shear stress for a simply supported beam with a uniform load is:

$$\text{Maximum shear} = w\ell/2$$

where w = Load on angle
= 40.8 lb/in.

ℓ = Length of opening
= 72 in.

Substituting these numbers into the equation gives the maximum shear of:

$$\text{Maximum shear} = V = w\ell/2 = (40.8)(72)/2 = 1469 \text{ lb}$$

The maximum shear stress, f, is determined using this shear of 1469 lb in the following equation:

$$f = 1.5V/(dt)$$

where V = maximum shear stress
= 1469 lb

d = length of vertical leg
= 4 in.

t = Thickness of leg
= 0.375 in.

When these values are used, the shear stress is:

$$f = (1.5)(1469)/(4 \times 0.375)$$
$$= 1469 \text{ lb/in.}^2$$

The maximum allowable shear stress for ASTM A36 steel is 14,400 lb/in.^{2}. This lintel easily satisfies shear stress criteria since the allowable shear stress of 14,400 psi is greater than the applied stress of 1469 lb/in.^{2}.

The maximum deflection for a simply supported beam is given by:

$$\text{Maximum deflection} = 5w\ell^4/(384\ EI)$$

where w = Load on angle
= 40.8 lb/in.

ℓ = Length of lintel
= 72 in.

E = Modulus of elasticity for steel
= 29,000,000 lb/in.^{2}

I = Moment of inertia
= 4.36 in.^{4}

When these values are used, the maximum deflection is:

$$\text{Maximum deflection} = \frac{5(40.8)(72)^4}{384\ (29{,}000{,}000)(4.36)}$$

$$= 0.11 \text{ in.}$$

This is less than the allowable of $\ell/600 = 0.12$ in.

Therefore, a 4 × 4 × 3/8 is satisfactory for the lintel shown in Figure 16-5. The summary of the design is as follows:

	Maximum	Allowable
Bending stress	17,393	24,000
Shear stress	1469	14,400
Deflection	0.11	0.12

Review of Table 16-2 verifies that the $4 \times 4 \times 3/8$ is adequate for supporting 12 ft of masonry.

The lintel to be designed for the next example is shown in Figure 16-6. Since the brick extends upward and sideways for a significant distance the masonry will arch across the opening. Because of arching the lintel only needs to support the loads shown in Figure 16-6.

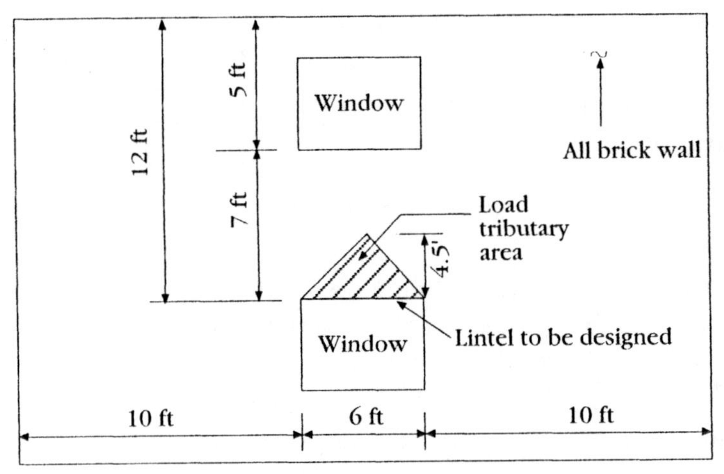

Load on lintel = 1/2 (width)(height)(weight)
= 1/2(6)(4.5)(40) = 540 lb

Figure 16-6—Lintel where arching does occur

The maximum bending moment stress for a triangular load on a simply supported beam is:

$$\text{Maximum bending moment} = W\ell/6$$

where W = Total load on angle
= Brick load plus estimated weight of steel
= 540 + 90
= 630 lb

ℓ = Width of opening
= 6 ft (72 in.)

Substituting these numbers into the equation gives the maximum bending moment of:

Maximum bending moment $= W\ell/6$
$= (630)(72)/6$
$= 7560$ in.-lb

The bending stress produced by this moment is determined by:

Bending stress = M/S

where M = Maximum bending moment
= 7560 in.-lb
S = 0.59 in.^{3} (assume $3 \times 3\text{-}1/2 \times 1/4$)

Substituting these numbers into the equation gives a bending stress of:

Bending stress = 7560/0.59
= 12,814 lb/in.^{2}

The maximum allowable bending stress for ASTM A36 steel under these specific loading conditions is two-thirds of the yield stress ($2/3 \times 36{,}000$ psi) = 24,000 psi > 12,814 psi. The lintel satisfies bending-moment requirements.

The maximum shear stress for this simply supported beam with a triangular loading is:

Maximum shear = W/2
where W = Total load
= 630 lb

Substituting these numbers into the equation gives the maximum shear stress of:

Maximum shear = W/2
= 630/2
= 315 lb

The maximum shear stress, f, is determined using this shear of 315 lb in the following equation:

$$f = 1.5V/(dt)$$

where V = Maximum shear
= 315 lb

d = Length of vertical leg
= 3 in.

t = Thickness of web
= 0.25 in.

When these values are used, the shear stress is:

$$f = (1.5)(315)/(3 \times 0.25) = 630 \text{ lb/in.}^2$$

The maximum allowable shear stress for ASTM A36 is 14,400 lb/in.^{2}. This lintel easily satisfies shear stress criteria.

The maximum deflection for this simple supported beam is given by:

$$\text{Maximum deflection} = W\ell^3/60EI$$

where W = Load on angle
= 630 lb

ℓ = Length of lintel
= 72 in.

E = Modulus of elasticity for steel
= 29,000,000 lb/in.^{2}

I = Moment of inertia
= 1.30 in.^{4}

When these values are used, the maximum deflection is:

$$\text{Maximum deflection} = \frac{(630)(72)^3}{(60)(29{,}000{,}000)(1.30)}$$

$$= 0.104 \text{ in.}$$

This is less than the allowable of $\ell/600$ = 0.12 in.

Therefore, the absence of the window allows for arching action and the angle can be reduced from a 4 × 4 × 3/8 to a 3 × 3-1/2 × 1/4.

	Maximum	Allowable
Bending stress	12,814	24,000
Shear stress	630	14,400
Deflection	0.10	0.12

Concrete Members

The portions of the structure that are in touch with the ground are usually concrete. The remaining three chapters in this text present three components of the structure that are made of concrete. Chapter 17 discusses concrete slabs which are used for the floor of the basement and garage. Slabs are also used for sidewalks and decks.

Chapter 18 presents information on the design of footings. Footings allow the load on the soil to be distributed over a greater area. This reduces the possibility of soil settlement.

Chapter 19 discusses concrete walls that are employed to retain the soil from spilling into the structure.

17

Concrete Slabs

Introduction

Concrete is the predominately used material for basement, garage and patio slabs. It is also used extensively in driveway construction. When properly placed, these slabs have extremely long service lives.

The thickness of these slabs are typically 4, 5 or 6 in. To obtain the expected service life, it is important to construct the slab on top of a properly prepared base. The base should be constructed of a granular material that has been compacted. If the material is not compacted, differential slab settlement might occur, leading to undesirable cracking. Failure to use a granular material might allow water to accumulate beneath the slab, causing upward heaving in cold weather.

Formwork for concrete slabs is relatively unsophisticated. Figure 17-1 depicts the details of the formwork used for a slab on grade. Formwork is often not needed since side containment is provided by existing walls and slabs directly adjacent to the slab that will be poured.

Although concrete is a versatile and durable material, it is susceptible to cracking. Such cracking is not only unsightly, it can allow moisture to penetrate the structure. For this reason care must be taken to reduce and control cracking.

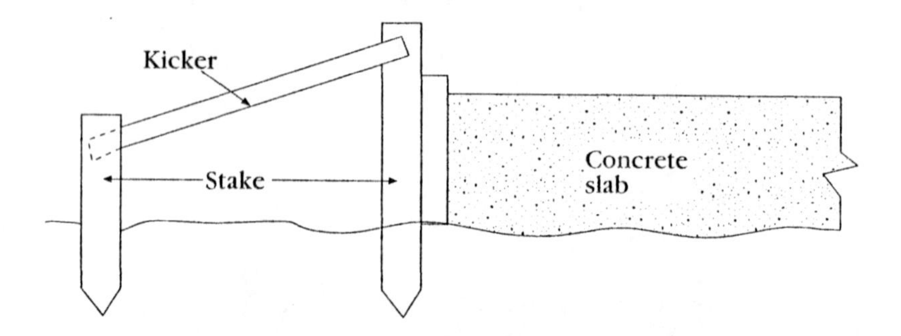

Figure 17-1–Side form for slab on grade

For the purpose of this discussion, concrete cracks are the result of the three following scenarios:

1) Overloading

2) Soil movement

3) Drying shrinkage

The first item, overloading, needs no further explanation since it is unlikely that a residential slab would be subjected to loads of any significant magnitude.

Soil movement can either be in the form of settlement or uplift, although the latter is rare. Soil settlement is caused either by compression of the soil from the weight of the concrete slab or by some internal repositioning of the soil particles. If the soil movement is not uniform, it will cause differential movement of the slab. This differential movement will induce stresses in the slab that might be large enough to lead to cracking. This same scenario occurs when the slab is subject to stresses from uplift forces. Slab uplift can be caused by swelling of the soil related to changes in water content of that soil. This water content change could be from a temporary rise in the water table or water from other sources (possibly a broken water or sewer pipe).

The third item listed is of particular interest since a considerable amount of cracking problems occur from this mechanism. As concrete dries, it will shrink in volume. This shrinkage will not necessarily result in cracking. Cracking, though, will occur when the shrinkage in the concrete is restrained.

Figure 17-2 depicts shrinkage of a concrete slab that is supported on a frictionless ventilated surface. Since the slab can contract freely and both top and bottom of the slab are drying at the same rate, no shrinkage cracking will occur. The shrinkage in the slab in inches due to the volume change is approximately equal the length of the slab in feet divided by 160. As an example, a slab that is 20-ft long will have 1/8 in. of shrinkage.

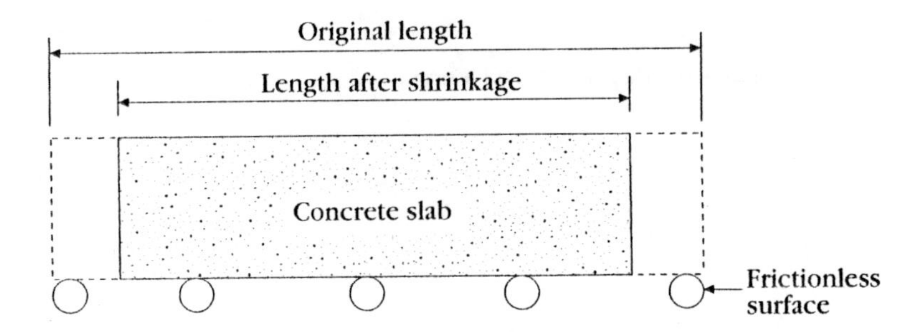

Figure 17-2–Shrinkage of slab on frictionless surface

Shrinkage cracking in slabs on grade is a combination of the concrete drying faster on the top of the slab in conjunction with restraint at the bottom of the slab. Since the top of the slab is exposed to air, it will tend to dry faster. This results in a volume reduction occurring faster at the top of the slab than at the bottom. The length of the top of the slab will shorten but the bottom of the slab will not cooperate with this movement. The movement in the top of the slab is further restrained by the friction of the gravel or sand base, preventing the bottom of the slab from sliding inward as it dries. Figure 17-3 depicts this scenario.

Shrinkage cracking can be reduced by proper curing of the slab. The use of steel reinforcement in the slab, as well as providing crack control joints, can reduce the crack width, the amount of cracking as well as control the crack location.

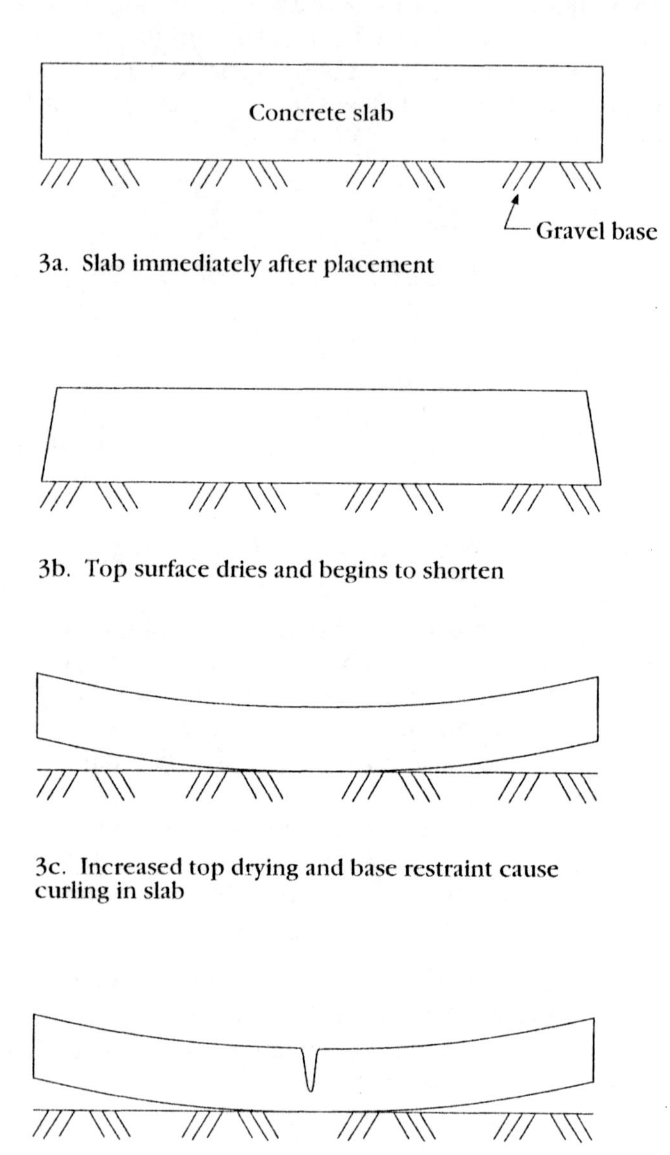

Figure 17-3—Cracking in concrete from shrinkage

Advanced Discussion

It is difficult to provide a slab on grade that is totally free from cracks. Care should be taken to minimize cracks since they are unsightly and may allow water penetration.

One of the most important aspects of placing an acceptable slab on grade is proper curing. Curing is the supplying or maintaining of moisture in the slab during the time immediately after placing the concrete until it has dried. Curing can be performed using one of three methods:

1. Provide surface moisture by producing a continuous spray of external moisture. A lawn sprinkler at a setting that produces a mist spray can be utilized for this method.

2. Placing wet burlap on top of the concrete or placing plastic sheets on the recently poured concrete to reduce the speed of evaporation of the moisture from the slab surface.

3. Reduce evaporation by spraying a chemical on the concrete that will retard the evaporation of moisture from the slab surface.

Keeping the concrete moist has several benefits. Concrete that is kept moist by curing will produce an end product that will have a faster gain of strength. This early strength gain will produce a concrete that is more resilient to drying induced cracking. Table 17-1 compares the rate of strength gain for concrete that was cured for 28 days, 3 days and 0 days. The early strength gains and final strengths are considerably higher when curing is utilized.

Although cracking is difficult to eliminate, the width of the cracks can be reduced by the use of steel reinforcing. Steel reinforcing is used in concrete members to work compositely with the concrete to produce a stronger and better-performing member. Steel placed in residential slabs on grade is not used for this purpose. Steel reinforcing in these slabs is typically used for controlling cracking associated with drying shrinkage and temperature changes. Steel reinforcement used in slabs on grade is typically welded wire fabric, although individual reinforcing bars placed in a criss-cross pattern can also be used.

Table 17-1—Strength Gain for 4-ksi Concrete with Various Curing Period

	Strength (No Curing)	Strength (3-Day Curing)	Strength (28-Day Curing)
Day 0	0	0	0
Day 3	1.3	1.6	1.6
Day 6	1.6	2.2	2.2
Day 10	2.0	2.8	2.9
Day 20	2.2	3.1	3.6
Day 28	2.2	3.2	4.0

Steel reinforcing does not eliminate cracking, rather, it constrains the width of the crack. Figure 17-4 depicts the benefits of providing reinforcing to reduce the widths of shrinkage cracks.

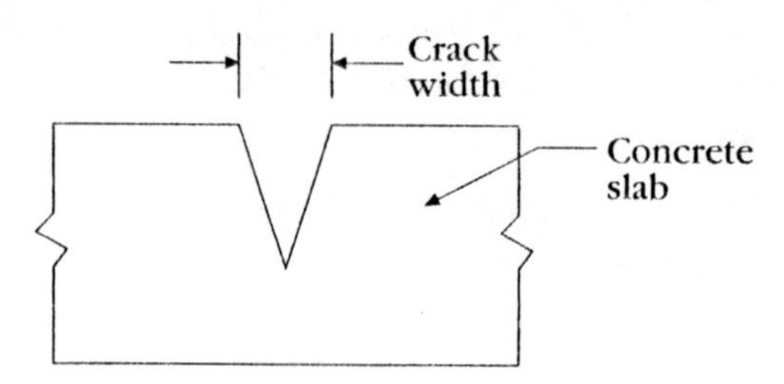

4a. Crack in slab without reinforcement

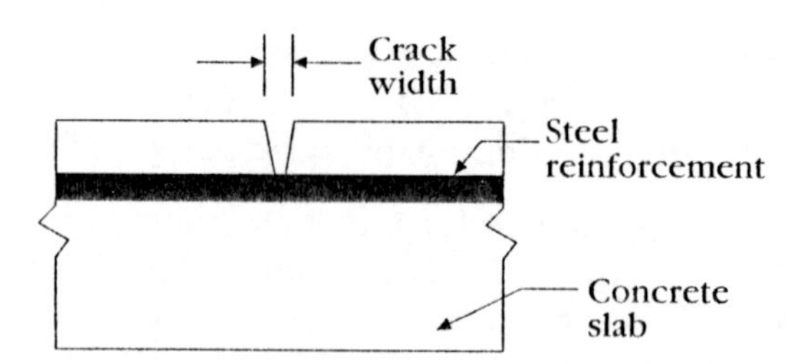

4b. Reduced crack width due to presence of steel reinforcement

Figure 17-4—Reinforcing for crack control

Concrete cracks tend to be wider at the surface. To reduce the crack width it is desirable to keep the reinforcing close to the top. Placing reinforcing too close to the top surface can also cause problems since it is closer to the elements that cause deterioration. It is suggested that the reinforcing be placed at a distance of one-third the slab thickness from the top of the slab. Figure 17-5 shows the detrimental effects of placing the reinforcing too low in the slab.

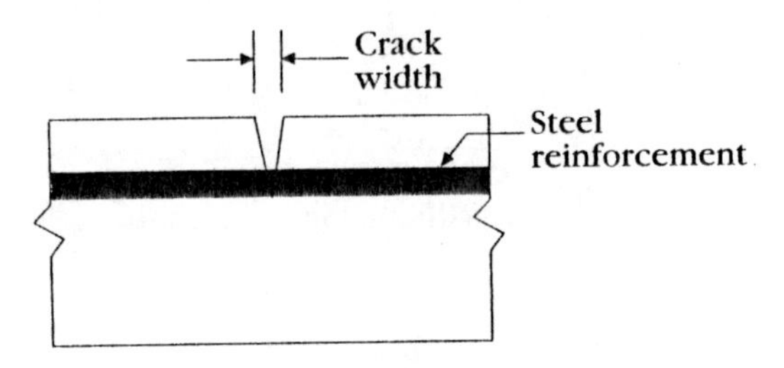

5a. Reinforcement near top of slab

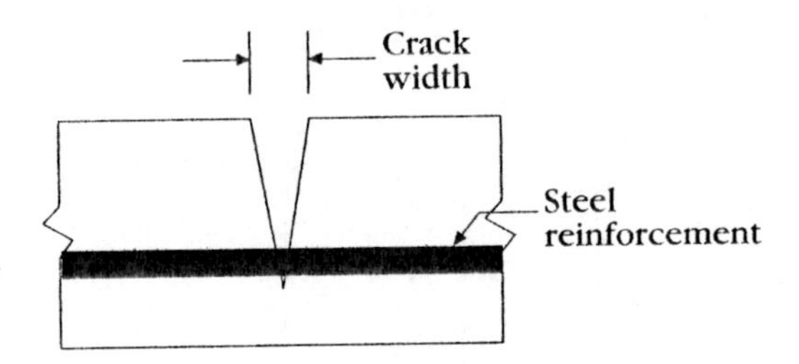

5b. Reinforcement near middle of slab

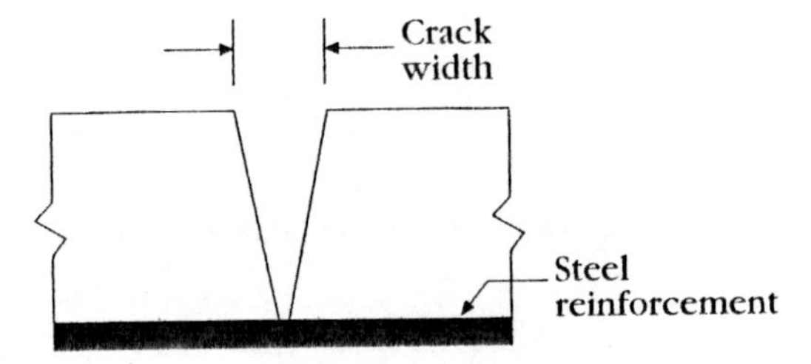

5c. Reinforcement near bottom of slab

Figure 17-5—Placement of steel reinforcing

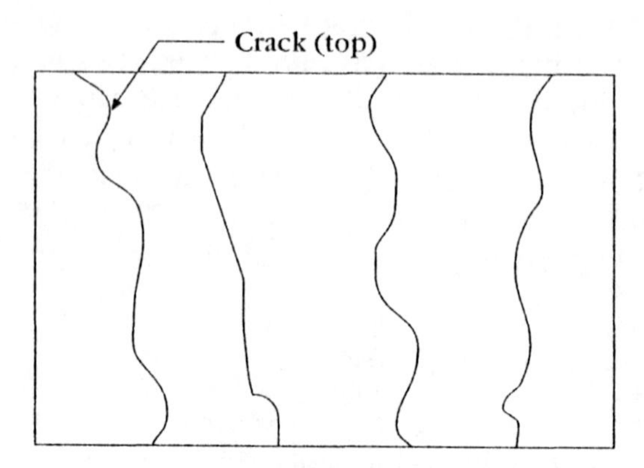

6a. Plan view of slab without any crack control joints

6b. Side view of slab without any crack control joints

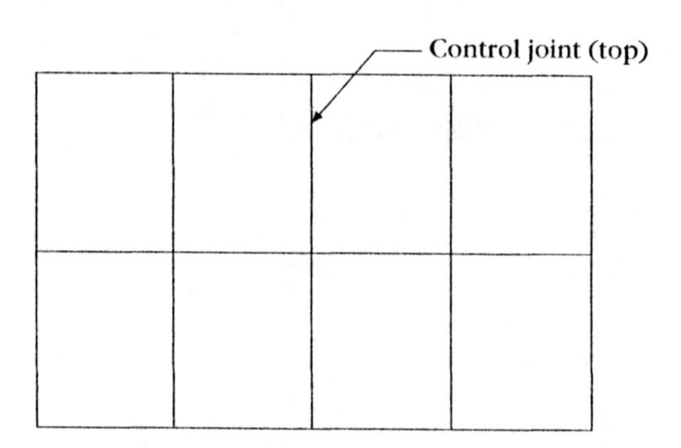

6c. Plan view of slab with crack control joints

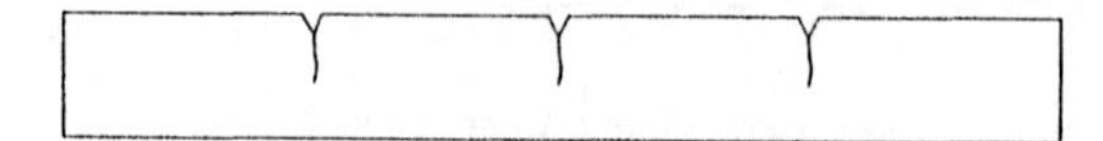

6d. Side view of slab without any control joints

Figure 17-6—Concrete control joints

Another method of reducing the detrimental effects of concrete cracking is the use of control joints. Crack control joints are preplanned grooves in the top surface of the concrete. These joints are quite apparent in every concrete sidewalk where they are 3 to 5 ft apart. As concrete shortens from shrinkage, the cracks will tend to locate themselves directly in the grooves. This is illustrated in Figure 17-6.

18

Concrete Footings

Introduction

Loads that are applied to any part of structure are transferred to the foundation, which in turn transfers the loads to the underlying soil. A foundation, also called the *substructure*, must be properly designed to transfer the loads from the structure above, called the *superstructure*, into the ground.

If the pressure on the soil is in excess of its bearing capacity, the structure will not be stable and damage or total collapse could occur. Typically the local building code will provide an allowable soil pressure for a given soil type. A given allowable pressure is not the load at which the soil will fail. It is the pressure that, if exceeded, could provide unsatisfactory performance results.

Estimating the bearing capacity at any given location is complicated. First, even soils that are classified as similar might have a wide range of bearing capacities. Second, no soil is a pure homogenous material. Not only is the soil a mixture of different materials, but it changes with depth. Strong layers of soil can surround a very poor-quality soil.

As an example, the allowable bearing capacities of different types of soils have values in the following ranges:

Sandy silts (loose)	1500 psf and under
Sandy clay (soft)	2000 psf and under

Sand (loose)	1500 psf to 6000 psf
Sand (compact)	3000 psf to 9000 psf
Clay (medium)	2000 psf to 6000 psf
Clay (hard)	4000 psf to 8000 psf
Gravel	9000 psf to 12,000 psf
Softrock	10,000 psf to 20,000 psf
Bedrock	40,000 to 200,000 psf

The residential substructure consists of three components. The first component is the column footing shown in Figure 18-1. The purpose of the column footing is to take the load at the base of the column and spread it over a larger area. This produces a smaller stress on the soil than would occur if the pipe were supported directly on the soil. The reduction in stress by spreading out the load is demonstrated in the following examples.

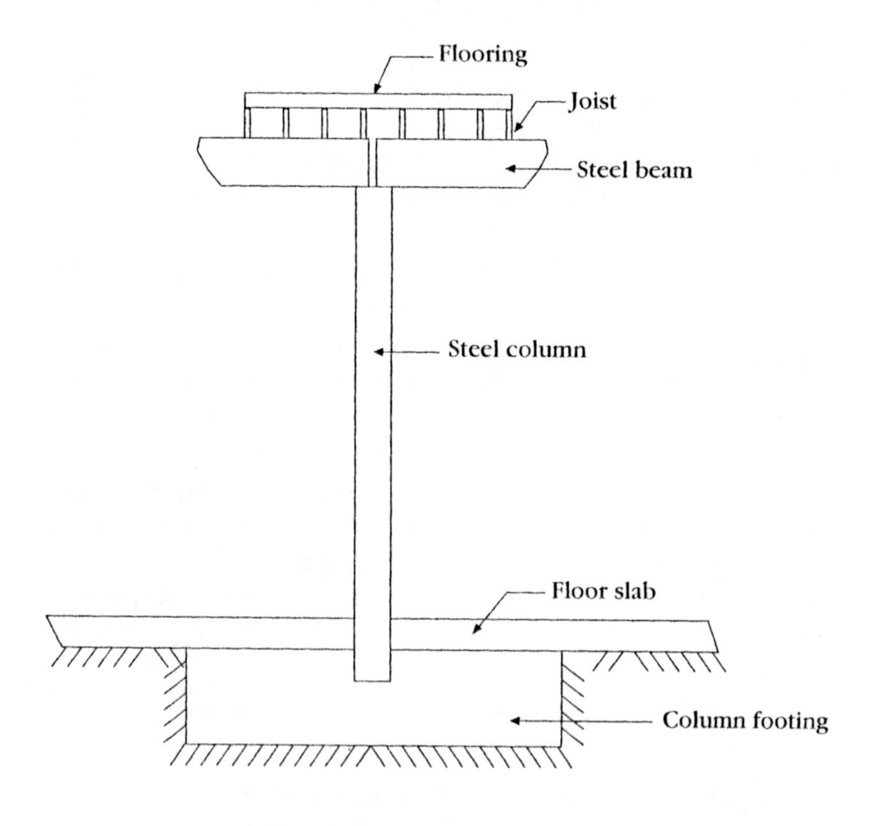

Figure 18-1—Column footing

Example 1: A solid column rests directly on the soil. The load on this 5-in. diameter column is 5000 lb. Determine the stress on the soil that results from the column loading.

$$\begin{aligned}\text{Bearing area of column on the soil} &= \pi D^2/4 \\ &= \pi \times (5)^2/4 \\ &= 19.6 \text{ in.} \\ &= 0.14 \text{ ft}^2\end{aligned}$$

$$\begin{aligned}\text{Pressure on soil} &= \text{Load/Bearing area} \\ &= 5000/0.14 \\ &= 35{,}714 \text{ lb/ft}^2\end{aligned}$$

Comparison of this value to the allowable bearing capacities previously listed reveals that this load would cause a local failure in the soil.

Example 2: A solid column rests directly on top of a concrete column footing. The load on the 5-in. column is 5000 lb. The size of the footing is 24 in. × 24 in. Determine the stress on the soil that results from the column loading on the footing.

$$\begin{aligned}\text{Bearing area of footing against the soil} &= \text{Length} \times \text{Width} \\ &= 24 \times 24 \\ &= 576 \text{ in.}^2 \\ &= 4.0 \text{ ft}^2\end{aligned}$$

$$\begin{aligned}\text{Pressure on soil} &= \text{Load/Bearing area} \\ &= 5000/4.0 \\ &= 1250 \text{ lb/ft}^2\end{aligned}$$

Comparison of this value to the allowable bearing capacities previously listed reveals that this load would produce bearing stresses low enough to allow bearing on the weaker soils.

The other two components of a residential substructure include the retaining wall and the wall footing (Figure 18-2). The retaining wall is typically a basement wall whose function is to contain the soil from falling into the basement.

The wall footing serves the same purpose as a column footing except the former is continuous along the length of the structure. The footing width is typically twice the width of the wall with a

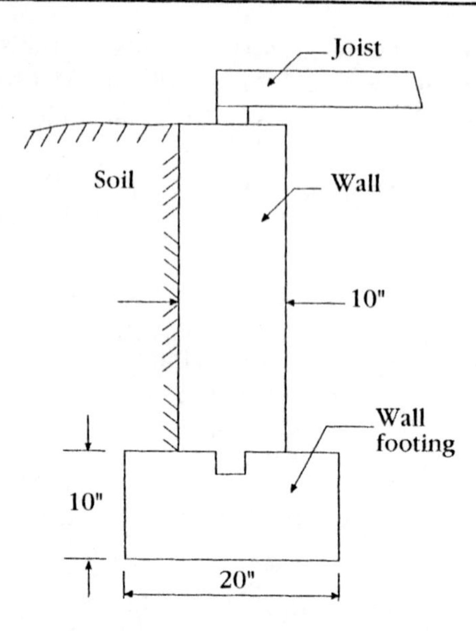

a. Basement wall and footing (no masonry)

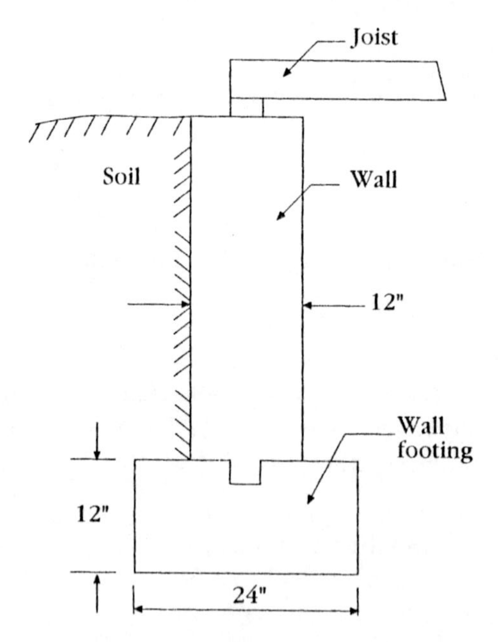

b. Basement wall and footing (masonry)

Figure 18-2—Basement walls

depth equal to the wall. Figure 18-2 shows typical dimensions for wall footings. The larger wall footing is used when the wall supports a brick facade. Since the brick dimensions are larger than siding thickness it requires that the wall be larger to support the brick.

The calculation of the bearing stress on the basement wall footing is calculated in a similar manner as the column footing.

Example 3: A concrete wall places a continuous load on a concrete footing of 1500 lb/linear ft. The wall does not support any masonry. Determine what the bearing stress is on the soil.

Since the wall does not support masonry, the footing is 20 in. wide. Therefore, the bearing for a 1-ft-wide section of a wall is calculated as:

$$\begin{aligned}\text{Bearing area of wall footing} &= 12 \times 20 \\ &= 240 \text{ in.} \\ &= 1.67 \text{ ft}^2/\text{ft}\end{aligned}$$

$$\begin{aligned}\text{Pressure on soil} &= \text{Load/Bearing area} \\ &= 1500/1.67 \\ &= 898 \text{ psf}\end{aligned}$$

Advanced Discussion

There are situations where the soil is of such a poor quality that the size of the footing is unrealistic. In these situations, the foundation must be supported on piles. These piles can be constructed of wood, steel or concrete. The piles transfer the loads of the structure by friction along its length or by bearing on the soil for a below-grade level.

A wood pile is made from a tree trunk with all of the branches removed. The wood pile has the advantage that it is not very costly to purchase. Its disadvantage is that it is hammered into the ground by the pile driver and also has limited lengths. A sketch of a wood pile is shown in Figure 18-3a.

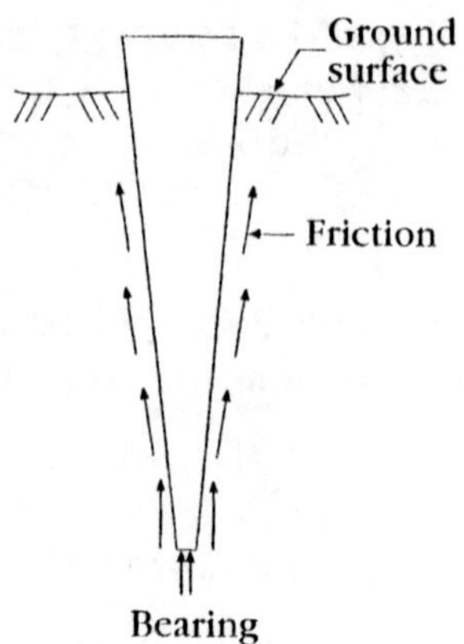

a. Wide pile

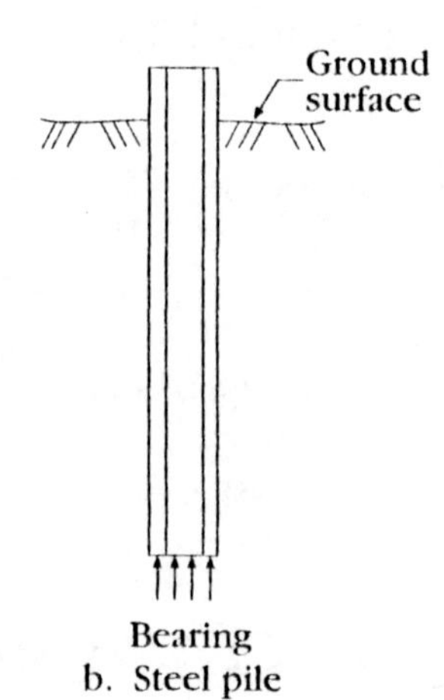

b. Steel pile

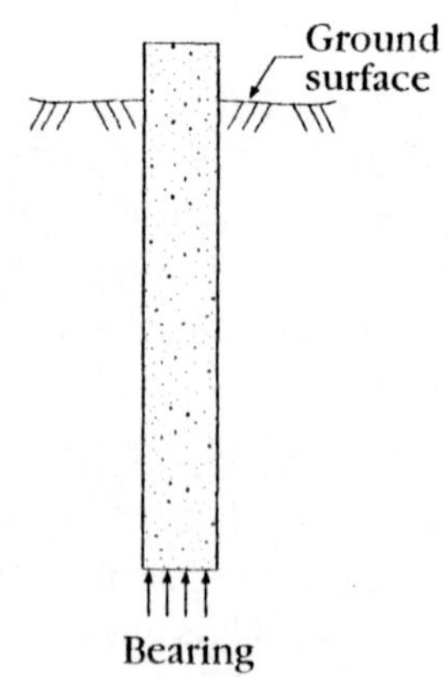

c. Concrete pile

Figure 18-3—Foundation piles

Steel piles have the advantage of being of unlimited lengths. As one steel member is hammered to its capacity, another member can be welded onto the end of the driven member. Steel piles have exceptional strengths but may be vulnerable to corrosion. A sketch of a steel pile is shown in Figure 18-3b.

Concrete piles, also called *caissons*, are usually not driven into ground. The foundations are constructed by digging out a hole and filling it with concrete. The advantage of concrete caissons is that they are economical. However, there are some soils where it may be difficult to drill the caisson hole. A concrete caisson is shown in Figure 18-3c.

19

Concrete Walls

Introduction

Concrete walls have two primary uses in residential construction. They can be used for soil retention and for basement walls.

Soil retention walls are needed when there is an abrupt change in the elevation of the ground. The concrete walls, called a *retaining wall* in this application, prevent the soil from falling on to the lower ground. If a wall is being used to retain soil of more than a few feet in height, the wall should be designed by an engineer. Figure 19-1 presents several different types of retaining walls. Figure 19-1a is the only one of the walls that would typically be utilized in residential construction.

The basement wall serves a similar function as the soil retention walls. However, the basement wall must also support loads from the structure above. The remainder of this chapter covers this type of concrete wall.

Figure 19-2 depicts a typical basement wall. The lower part of this detail shows a wall footing, which is discussed in the previous chapter. This figure also shows a portion of the wall that fits into the footing. This slot is called a *keyway*. A keyway locks the wall into the footing so that the wall will not slide inward when the soil loads are applied.

Walls are constructed using the formwork shown in Figure 19-3.

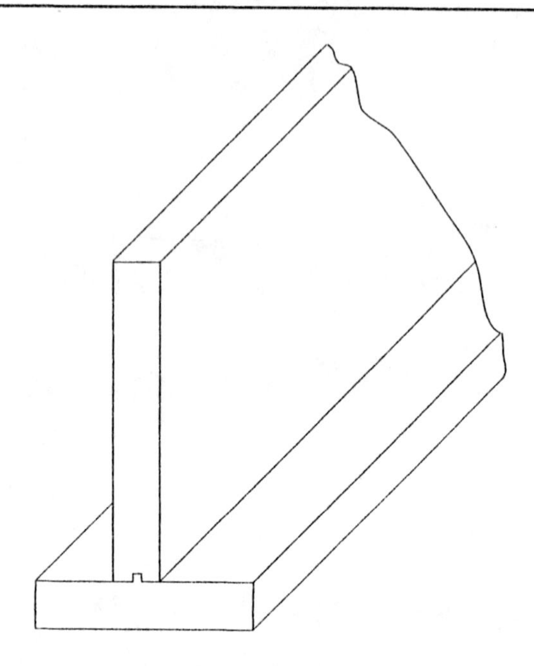

1a. Retaining wall

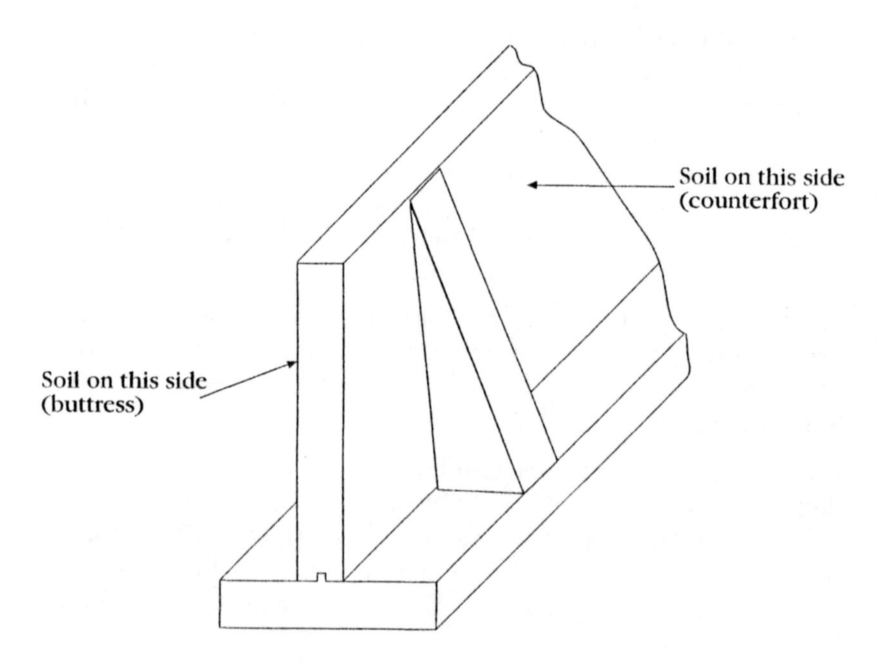

1b. Counterfort or buttress wall

Figure 19-1—Soil retaining walls

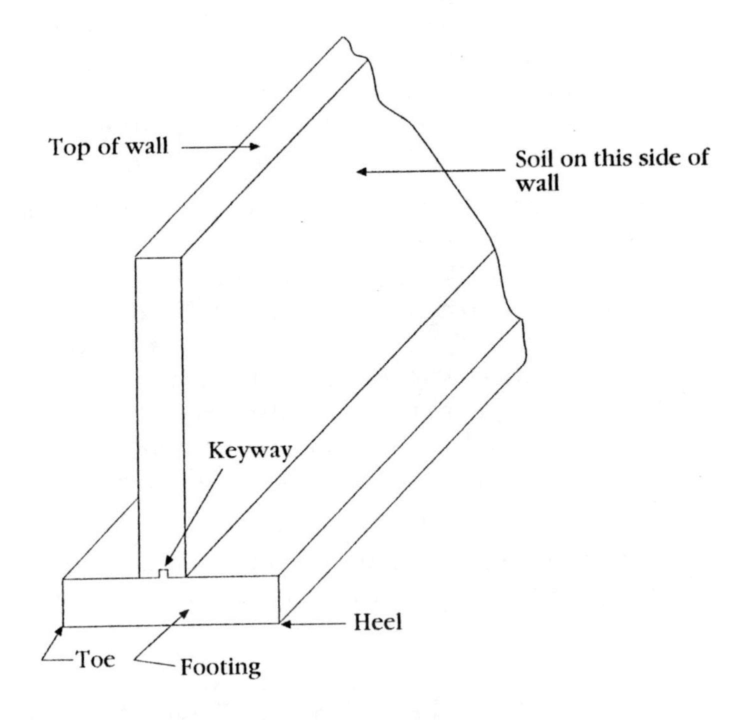

Figure 19-2—Basement wall

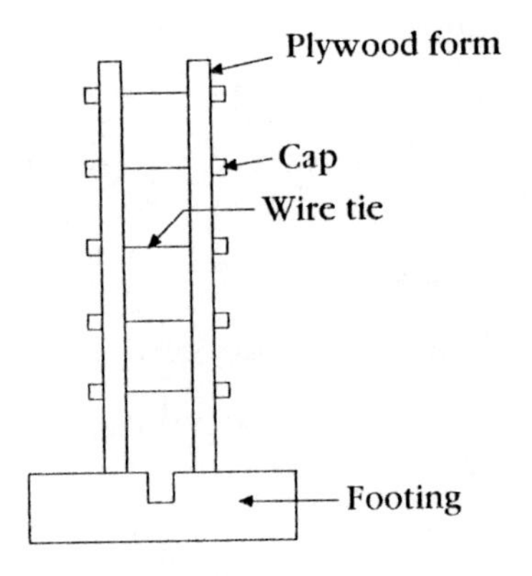

Figure 19-3—Wall formwork

The components of the formwork are the plywood sheets, which contain the concrete and steel wire ties that prevent the formwork from spreading outward when filled with concrete. The plywood sheets must be coated with a substance that will prevent the plywood from bonding to the concrete. This will provide for a clean removal of the plywood so that it can be reused.

The force that the soil exerts against the wall is a function of the soil type and the stiffness of the wall. Determining this force can be complex. Building codes simplify this matter by requiring that the wall be designed for a given lateral load. For example, 40 lb/ft^2 per foot of height. That is, the first foot down of the wall must be designed to resist 40 lb/ft^2, 2 ft from the top must be designed for 80 lb/ft^2, 3 ft from the top must be designed for 120 lb/ft^2, etc. It can be seen that the bottom of the wall will be subjected to a large lateral load. At 9 ft below grade, the lateral force is 360 lb/ft^2.

The force on the wall may be increased by nearby loads on the ground surface. These loads can be from machinery or piles of soil.

Advanced Discussion

This portion of the chapter presents calculations for the stress in concrete basement wall.

Example: Determine the maximum stress in a concrete wall that is used for a basement in a residential structure.

Assume the following:

- Wall is 8 ft high
- Lateral pressure produced by soil against the wall is 40 lb/ft of height of wall.
- Wall is 10 in. thick
- Basement slab and first-floor joists have been installed.

Under these conditions, the wall has the loading conditions shown in the following.

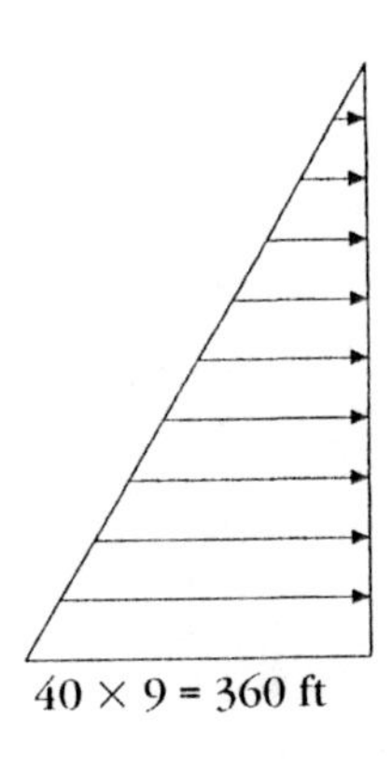

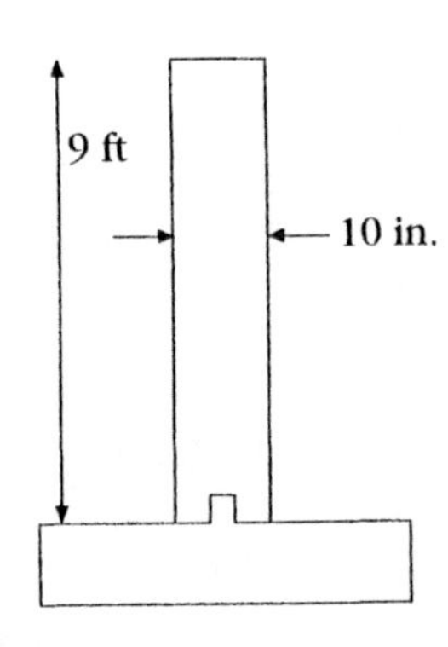

Under the triangular loading condition, the maximum bending moment is:

$$M = w\ell^2/15.5$$

where: w = Distributed load at base of wall
= 360 lb/ft
= 30 lb/in.

ℓ = Height of wall
= 8 ft
= 96 in.

Substituting those numbers into the equation gives the maximum bending moment:

$$M = (30)(96)^2/15.5$$
$$= 17{,}837 \text{ in.-lb}$$

This produces a stress as follows:

Maximum stress = Md/2I

where: M = Maximum bending moment
= 17,837 in.-lb

d = wall thickness
= 10 in.

$$
\begin{aligned}
I &= \text{Moment of inertia} \\
&= bd^3/12 \\
&= (\text{1-ft strip})(\text{thickness})^3/12 \\
&= (12)(10)^3/12 \\
&= 1000 \text{ in.}^4
\end{aligned}
$$

$$\text{Maximum stress} = \frac{(17{,}837)(10)}{2 \times (1000)}$$

$$= 89 \text{ psi}$$

This tensile strength in the concrete should be great enough that cracking will not occur if the wall is allowed to gain strength prior to backfill being added.

Glossary

Additives–Chemicals placed in the concrete mixture to improve one or more qualities of the concrete. The additives might be used to aid in the placement process, such as drying accelerators or additives that make the concrete more liquid. Additives can also be for the final product, such as for color pigmentation.

Air Content–Concrete that is subjected to cycles of freezing and thawing might begin to crumble on its surfaces. This problem can be avoided by introducing air bubbles into the concrete mixture. As the water in the edges of the concrete freezes, it will expand into the air pockets rather than build up pressure and break the top layer of concrete.

Allowable Span Length–The maximum distance that a beam can traverse before it exceeds code prescribed deflection or stress limitations.

Anisotropic–A member is said to be anisotropic when its properties in all three directions (length, width and depth) are different. Wood is an anisotropic member since its strength is different, depending on which way it is oriented.

Arching–The beam over the top of an opening must support the loads applied above the opening. These loads are less than one would expect since material, particularly masonry, has the capability of behaving similar to an arch. It will tend to thrust its loads around the opening.

ASTM–Abbreviation for the American Society of Testing Materials. This society provides specifications for all types of construction materials and their components.

Base Preparation–Material that is directly below a concrete slab is called *base material.* If the base material is improperly placed, the concrete slab will experience differential settling and cracking. Proper base preparation includes placing aggregate and compacting it prior to placing the slab.

Bending Moment–A twisting force on a beam or column that is produced by the applied loads. The applied loads cause these members to develop curvature. This curvature produces a bending moment. Bending moments cause the top fibers to shorten and the bottom fibers to elongate if the beam is bowing downward. The shortening produces compression and the elongation produces tension.

BOCA–Abbreviation for Building Official Code Administrators. It typically refers to the building code that is published by this association.

Bolts–Connectors used in residential construction for the purpose of tying beams to columns, and the walls to the foundation.

Box Nails–A type of nail that is often used for connecting framing members in residential construction.

Bridging–Short members that are placed diagonally between parallel joists. These short members brace the joist to the adjacent joist to assist the floor system in acting as a unit rather than a series of single members.

CABO–Building code that is used for one-and two-story dwellings. This code represents a compilation of requirements from the major code writing bodies.

Calibrated Wrench–A tool used for tightening bolts to a predetermined tension. This wrench is similar to an ordinary wrench except that it will not operate beyond a predetermined torque.

Cantilever–A beam that has one end restrained from any type of movement and another end that is totally free to move.

Channel–A steel member that has a cross section that has a shape similar to the letter "C."

Chord–The perimeter members of a truss. These members are typically horizontal or are on a slight slope. Chords are classified as top or bottom chords. Top chords are usually compressed and bottom chords tensioned when loads are applied.

Codes–Regulations that provide the requirements for all phases of construction, including plumbing, electrical and construction. A jurisdiction will adopt a code that will provide the rules that must be satisfied to obtain an occupancy permit. The three major building codes in the United States are UBC (Uniform Building Code), BOCA (Building Officials Code Administrators) and SBC (Standard Building Code).

Common Nails–A type of nail that is often used for connecting framing members in residential construction.

Compression–The result from a load that causes a member to shorten. If the load causes the member to lengthen, it is in tension rather than compression.

Concentrated Load–A force that is applied over a very small area. For analysis purposes, the load is considered to be applied at a distinct point.

Concrete–A mixture of water, stones, and cement that hardens into a rock-like mass with superior load-carrying abilities.

Concrete Types–Concrete mixtures are available in various types for various uses. For instance, Type 2 concrete is for structures that are under water. Other types are used when high strength is needed or when the environmental conditions warrant special considerations.

Continuous Beams–A structural member that has more than two supports. In residential construction, a beam with only two supports is called *simply supported.*

Control Joints–Grooves placed in concrete to establish the location where cracking should occur. These joints do not reduce cracking but look less unsightly because the cracks tend to be located in the grooves.

Curing–Denotes the time period immediately after concrete is placed and the concrete begins to dry. Controlling the rate of drying could possibly be crucial to its future performance.

Dead Load–The load that is placed upon a structure that is the result of its own weight.

Direct Tension Indicator–A measuring device used to determine the amount of tightening that has occurred in a bolt.

Drift Load–A structure is designed for a given snow load, based upon its locale and other factors. In some cases, the additional snow might accumulate because of wind-blown snow collecting on the roof. This additional load is called a *drift load.*

Drywall–The wood frame of a structure needs an architectural finish to cover the exposed members. A preformed gypsum product ranging from 3/8 in. to 3/4 in. is the most popular material for this use. Prior to the availability of drywall, the most popular finish was lathe and plaster.

Early Wood–The portion of the early growth of a tree that occurs in the beginning of the growing seasons. The amount of growth that occurs as early wood and late wood will determine how dense the wood will be.

Essential Facility–A distinction in building codes that designates what buildings are of great importance in the event of a disaster. These buildings must be designed for larger loads than other buildings. The purpose of this is to ensure that these designated buildings are available and functioning after a local disastrous event.

Exposure Classification–A designation that denotes the environment that the materials will be subjected to when in service. These designations assist in the selections of a material that is appropriate for the specific conditions the material will experience.

Facade–Exterior covering of a structure. Facades are typically brick, vinyl siding, aluminum siding, cedar siding and stucco.

Fixed Support–Columns and walls hold up beams. When these members that hold the beams prevent displacement and rotation it is considered fixed. If the support only prevents displacement and allows the beam to rotate freely, it is a simple support.

Flange–Denotes the top and bottom of a cross-section of a structural member. As an example, in an I beam, the horizontal portions of the cross-section are flanges. The vertical part is the web.

Flitch Plate–A wood beam has its strength limitations and may not be able to satisfy the requirements of a given loading condition. Several beams can be put side by side in this case. However, in some cases, the multiple-beam solution is still inadequate. Under these circumstances a steel plate can be attached to the wood beam to increase its strength. This steel is called a *flitch plate*.

Floor Covering–An architectural finish placed on top of the plywood decking. Common floor coverings are ceramic tile, vinyl, carpet and wood.

Footing–The wide portion of a wall that rests directly on the soil. The footing is designed to be a width that will appropriately distribute the soil load to avoid any overloading of the soil.

Formwork–Mold into which concrete is placed to allow the concrete to harden into the desired shape at the desired location. Formwork is constructed with specially treated plywood. The plywood is coated with liquid that prevents the concrete from sticking to the wood and therefore allows for its easy removal.

Glued Laminated Beams–A structural member that is fabricated by gluing together long, thin pieces of wood. Glued laminated beams provide superior strength since the plies of wood are only used if they are of superior quality. Also, the glue has strength in excess of that of the wood, preventing premature failure.

Hardwood–Dense wood used for making cabinets, trim and furniture. Dense wood typically comes from trees that have leaves rather than needles.

Header–A deep beam that carries the loads above an opening in a wall to the sides of the wall. Headers must be able to support a substantial load without having excessive deflections to prevent jambing of windows and doors as well as cracking of plaster and veneer.

High-Strength Bolts–Bolts that are utilized to connect steel structures. The chemistry of the bolts provides a strength that exceeds the strength of common bolts.

Importance Factor–A numerical factor that increases the forces applied to a structure that is directly related to the type of structure being designed. Essential facilities such as hospitals and police stations have an importance factor larger that 1.0 to provide a structure with a larger strength reserve than typical structures.

Joist–Beams that are spaced close together to form a structural system used to support a floor or roof. Joists can be wood, steel or concrete. Wood joists are typically spaced at 16 in. for residential construction.

Joist Spacing–The distance of the center of a joist to the center of the adjacent joist.

Lintel–Steel members that are placed at the top of a window or other opening. The weight of the bricks above the opening are supported by the lintels. The lintel, typically a steel angle, carries the loads to the side of the opening.

Live Load–Nonpermanent loads that are caused by the occupants of a structure. These transient loads are usually associated with the weight of people.

Live-Load Reduction–Structural members must be designed to carry the live load prescribed by the code on the entire floor area that the member supports. In reality it would be unlikely that the entire floor area would be loaded to the maximum at the same time. Building codes recognize this impossibility and provide a reduction factor.

Load Duration Factor–A modification of the allowable load carrying capacity of a beam for wood beams that have long-term loads. This reduction in load carrying capacity recognizes that wood beams under sustained loads have lower strengths than those beams under temporary loads.

Local Terrain–The topography of the area where a structure will be or is located. Local terrain has a direct bearing on the intensity of wind forces that can develop.

Low-Hazard Facility–A structure that is not likely to injure people if it fails because the structure has little chance of being inhabited.

Maximum Bending Moment–The forces in beams and columns vary along the length of the members. For design purposes the largest value is critical. The maximum bending moment is the value used to design the member for stress.

Maximum Shear–The forces in beams and columns vary along the length of the members. For design purposes the largest value is critical. The maximum shear is the value used to design the member.

Modulus of Elasticity–Provides a numerical relationship between the change in length of a member and the force it takes to produce that change. The modulus of elasticity is large for materials that deform easily such as rubber and is smaller for materials that do not, such as concrete.

Mortar–A mixture of cement and water that is used to glue adjacent bricks to one another to form a monolithic wall.

Petrography–The study of concrete by viewing and testing on a microscopic level similar to the way one would study rocks.

Ply–One of several thin layers of wood glued together to form plywood.

Plywood–Thin layers of wood, called *plys* are glued together to form this material. Each layer is rotated to have its grain perpendicular to the adjacent layer to provide a member with significant strength. Plywood typically comes in 4 ft × 8 ft sheets.

Plywood Decking–The use of plywood as a floor or roof surface. The plywood transfers the applied loads to the beams and joists while providing a flat surface.

Popout–Defect on the surface of concrete that is caused by the aggregate expanding and breaking off the finished surface of the concrete.

Post Tensional Concrete–A concrete member that is prestressed after the concrete has hardened.

Precast Concrete–Concrete members that are constructed at a location other than where the member is to be placed for its final position. These members are constructed off site, shipped to site, lifted into place and connected.

Prestressed Concrete–Inducing stresses in a concrete beam that are opposite to what the member will experience during its use. A beam that will deflect downward under loads can be prestressed such that it bows upward prior to the addition of the loads. After the loads are added, the bowing will offset the downward deflection, resulting in a member that has no stresses.

Pretensioned–A concrete member that is prestressed before the concrete hardens.

Roof Loads–Forces, usually snow and wind, applied to the top of a structure.

Rowlock–Describes the side that a brick has been laid on. Rowlock is when the brick is resting on its shortest dimension.

SBC–Abbreviation for the Standard Building Code. This code typically applies to the southwest portions of the United States.

Scaling–Flaking off the top surface of concrete. Usually caused by excessive troweling of the top surface during placement.

Serviceability Requirements–The components of a structure must be designed to have strength that is in excess of the forces that the loading will produce. This strength requirement does not guarantee that the structure will be satisfactory to the occupants. The structure must be able to resist the forces while not having excessive deflections or cracking. These additional requirements are called *serviceability requirements*.

Setting Accelerators–A concrete admixture used to speed up the time it takes for concrete to harden.

Shear–One of the internal forces produced by the applied loads. This force is responsible for the cross-section of the beam trying to slip past the adjacent cross-section.

Shear Key–A continuous notch that is placed in the top of the footing where the bottom of the wall is poured. The shear key prevents the wall from sliding off the footing.

Shrinkage–The reduction in volume of concrete as it dries.

Simple Support–Columns and walls hold up beams. When these members that hold the beams prevent displacement but not rotations, the support is considered a *simple support.*

Sliding Snow–Roof structures are designed to carry snow loads that have fallen from the sky. In some cases the snow falling from higher buildings may increase the load on a lower adjacent building. This extra snow load is called *sliding snow.*

Slump Test–A method used on construction sites for measuring the stiffness or fluid tendencies of a batch of wet concrete.

Soft Wood–Wood that is used for forming the structural frame of a building. Softwoods usually come from trees that have pine needles.

Steel Angles–Metal beam that has a cross-section that is in the shape of an "L." Steel angles are used to carry the weight of bricks over an opening in the structure.

Steel Reinforcing–Metal bars placed inside concrete to assist the concrete in carrying the applied loads. Concrete is weak in tension. The addition of the steel reinforcing will vastly increase the tensile capacity of the concrete member.

Strength Requirements–Stipulation that loads do not induce stresses in excess of what the allowable stresses are in the code.

Studs–Vertical members spaced 12 to 24 in. apart (usually 16 in.), that carry the horizontal and vertical loads imposed on the structure.

Support–A member that holds up another member. Concrete walls would be considered the supports for a wood joist that rests on it. The exterior walls would be the support for the roof trusses.

Surcharge Load–A force applied at the ground surface that results in pressures in the soil below.

Tension–The results from a load that causes a member to become longer in length. If the load causes a member to shorten, it is in compression, rather than tension.

Tributary Area–The portions of the floor that will contribute loads to a particular member.

UBC–Abbreviation for the Uniform Building Code. This code applies to the West.

Uniformly Distributed Load–Forces on a member that are equal for the entire length. Loads that are spread out evenly over the tributary area.

Veneer–The architectural covering of the structural frame of a building. The architectural covering is typically brick, aluminum siding, vinyl siding, cedar siding or stucco.

Welding–The fusing together of two pieces of steel by heating.

Wood Grades–The division of lumber into different categories, based on quality characteristics.

Yield Strength–As a force is applied to a material, it will either shorten or elongate, based on the direction of the load. As the load increases, the material will continue to shorten or elongate. At some time the load will reach a point where the member will continue to change in length without the addition of anymore load. This point is the yield strength.

Index

About the Author

Dr. August W. Domel, Jr., is a licensed structural engineer, professional engineer and attorney at law in Illinois specializing in failure investigation and construction law. Currently he is Adjunct Assistant Professor at the Illinois Institute of Technology (I.I.T.) teaching construction management and construction law courses and manager of structural engineering at Engineering Systems, Inc. in Aurora, Illinois.

He graduated Summa Cum Laude from Bradley University with a bachelor of science degree in civil engineering. He received a master's degree and Ph.D. in civil engineering from I.I.T. and the University of Illinois at Chicago, respectively. He also earned a law degree from Loyola University of Chicago.

He has written books on the topics of earthquake design of high-rise buildings, design of water-retaining structures, concrete floor design and estimating. Most recently he co-authored *Residential Contracting—Hands-on Project Management for Builders* and *Legal Manual for Residential Construction* published by McGraw-Hill.